CIS-TRANS
ISOMERIC CAROTENOIDS
VITAMINS A
AND ARYLPOLYENES

BY

L. ZECHMEISTER

CALIFORNIA INSTITUTE OF TECHNOLOGY
PASADENA, CALIFORNIA

WITH 112 FIGURES

1962

NEW YORK
ACADEMIC PRESS INC. PUBLISHERS

VIENNA
SPRINGER-VERLAG

SPRINGER-VERLAG
VIENNA

Published in USA and Canada by

ACADEMIC PRESS INC., PUBLISHERS
111 Fifth Avenue, New York 3, New York

ISBN-13: 978-3-7091-5550-9   e-ISBN-13: 978-3-7091-5548-6
DOI: 10.1007/978-3-7091-5548-6

TO ELIZABETH

Preface.

The core of this little book is a survey of the experimental work carried out by our research group in the course of two decades. Although the writer has made an effort to present and evaluate related contributions of many authors, the text should not be considered as a monograph of encyclopedic character.

Our experiments began with a study of *cis-trans*-isomeric C_{40}-carotenoids and were subsequently extended to other types of polyenes in order to determine how far the observations were applicable to broader fields.

Similar considerations have led to the inclusion of the Chapters on Vitamins A and on Cumulenes although we have not contributed to the stereochemistry of these classes of compounds.

The first eleven Chapters have been published in Volume XVIII of the Series, "Progress in the Chemistry of Organic Natural Products" and now appear in a revised, up-to-date form.

Sincere thanks are due to the following former members of our research group who have participated in the experimental elucidation of the field: W. V. Bush, A. Chatterjee, L. Cholnoky, J. Dale, R. B. Escue, P. Fischer Jörgensen, Ch. Gansser, F. T. Haxo, G. Karmakar, B. K. Koe, J. Leemann, R. M. Lemmon, the late A. L. LeRosen, W. Lijinsky, K. Lunde, E. F. Magoon, W. H. McNeely, F. J. Petracek, J. H. Pinckard, A. Polgár, A. Sandoval, W. A. Schroeder, J. W. Sease, Y. W. Tang, K. Tsukida, the late P. Tuzson, L. Wallcave, and B. Wille. Dr. Bush has kindly read the manuscript and the galleys.

The reader will note that much of the material presented on the following pages is of empirical nature. It is hoped, however, that younger investigators who have newer and more refined experimental tools at their disposal will develop theoretical explanations for many of the recorded phenomena, physical, chemical and biological.

Pasadena, Autumn 1961. L. Z.

Contents.

List of Figures.

Page

First Part

Isoprenic Polyenes.

I. Carotenoids: Introduction.

As is well known, the carotenoid pigments belong to the class of polyenes and represent the most unsaturated mass products of biosynthesis. Several carotenoid pigments are extremely widespread in the vegetable and animal kingdoms. Although the name "poly-ene" would a priori include any compound whose molecules contain more than two carbon-carbon double bonds, the accepted use of the term is restricted to long, aliphatic, conjugated double bond systems.

The carbon skeleton of the carotenoids is highly branched and composed of isoprenic building blocks. Most naturally occurring carotenoids contain 40 carbon atoms, corresponding formalistically to 8 isoprene units. The number of conjugated $C=C$ double bonds varies from 7 to 15 and is 10 or 11 in many common carotenoids. By synthesis, analogous systems up to 19 conjugated double bonds have been obtained (*228*).

The plant carotenoids show intense yellow, orange, red, or violet color. They are accompanied in the tissue by "colorless carotenoids" such as phytofluene (*527, 510, 417, 368*) in which only 5 double bonds are conjugated. Phytofluene (p. 102) is practically colorless but fluoresces intensely in ultraviolet light. The lowest member of this series is phytoene (p. 102) (*370, 379, 378*) with 3 conjugated double bonds; it shows neither color nor fluorescence. The two colorless compounds mentioned have the same carbon skeleton as the colored ones and can be dehydrogenated in vitro to carotenoid pigments (*511, 499, 249*).

It is a general feature of the carotenoids that their chromophores do not extend over the whole main carbon chain, although they occupy the largest, middle section of the molecule. The latter carries two characteristic "terminal groups" that have either aliphatic or hydro-aromatic or (exceptionally) aromatic structure. In a number of common carotenoids α- or β-ionone rings are found at the molecule ends. As an exceptional feature the paprika ketones capsanthin and capsorubin possess cyclopentane end groups (p. 95).

Since the two terminal groups may be either identical or different, both symmetrical and non-symmetrical carotenoid structures are known.

The number of the more or less well-characterized, naturally occurring carotenoids is about 80, to which several totally-synthetic products which do not occur in plants or animals have been added in recent years.

In considering functional group content the carotenoid pigments include hydrocarbons, alcohols, aldehydes, ketones, carboxylic acids, epoxides, furanoid oxides, and combinations of these types. The O-function is located in terminal groups and not in the multiconjugated system. Evidently, this feature as well as the presence of characteristically arranged methyl side-chains is determined by the path of biosynthesis.

In some instances a C_{40}-carotenoid undergoes partial oxidative cleavage in the plant to give lower-molecular weight pigments carrying either aldehyde or carboxyl groups at the point(s) of attack.

Examples. β-Citraurin (C_{30}) (*533*) and β-apo-8'-carotenal (C_{30}) (*481 a*); for some C_{20}-and C_{24}-dicarboxylic acids cf. Chapter X, p. 107.

In most C_{40}-carotenoids such as α-carotene, β-carotene, their hydroxylated derivatives, etc., the hydroaromatic rings and the middle section of the molecule are held together by means of single bonds; this defines a *normal* (cyclohexenyl) structure (V). In less frequent instances, double bonds play this part and then the term *retro* (cyclohexylidene) structure is used (*337*) (VI, p. 6).

A detailed discussion of the biosynthesis, occurrence, isolation, structural clarification and in vitro synthesis of the carotenoids lies outside the scope of the present stereochemical survey, and the reader is referred to pertinent monographs (*491, 232, 135, 210*). Certain in vitro conversions of natural carotenoids have been surveyed recently (*499*).

The structural formulas of some important carotenoids are listed on pp. 5–7; several other structures will appear later.

The conventional manner of writing carotenoid structures is shown in the formulas (I)–(IV), while in (V) and (VI) only carbon skeletons appear. Finally, the abbreviated symbols (VII)–(XXIII) are restricted to the terminal groups; in such formulas the dots represent the following grouping:

Nomenclature.

As shown in the β-carotene formula (I), the C-atoms of the main chain are numbered 1–15 and 1'–15', beginning at the carbons which carry *gem.* methyl groups (KARRER's nomenclature). In the presence of two different hydroaromatic rings, the β-ionone ring carries the unprimed numbers [example, (II)]; and when one of the terminal groups is an open chain, the latter is primed [example, (III)]. In the case of fully aliphatic carotenoids the number 1 and 1' will be assigned to the same carbon atoms as in the corresponding cyclic structures [example, (IV)]. The latter arrangement, of course, does not comply with assigning

(I.) β-Carotene.

(II.) α-Carotene.

(III.) γ-Carotene.

(IV.) Lycopene.

No. 1 to the first carbon of the longest chain according to the Geneva Nomenclature. ·

The term "stereoisomeric set" includes all possible *cis-trans* forms of a given carotenoid, and each stereoisomer is a "member" of the set.

(V.) β-Carotene *(normal)*.

(VI.) *retro*-Dehydrocarotene.

(VII.) β-Carotene.

(VIII.) Cryptoxanthin (3-hydroxy-β-carotene).

(IX.) Isocryptoxanthin (4-hydroxy-β-carotene).

(X.) Zeaxanthin (3,3'-dihydroxy-β-carotene)

(XI.) Isozeaxanthin (4,4'-dihydroxy-β-carotene).

(XII.) Canthaxanthin (4,4'-diketo-β-carotene).

(XIII.) β-Carotene monoepoxide.

(XIV.) β-Carotene diepoxide.

(XV.) 3,4-Dehydro-β-carotene.

(XVI.) 3,4,3',4'-Bisdehydro-β-carotene.

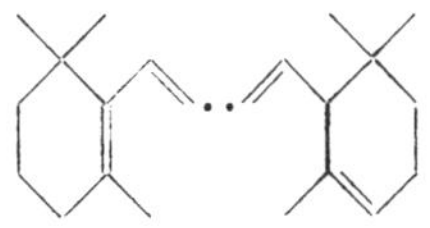

(XVII.) α-Carotene.

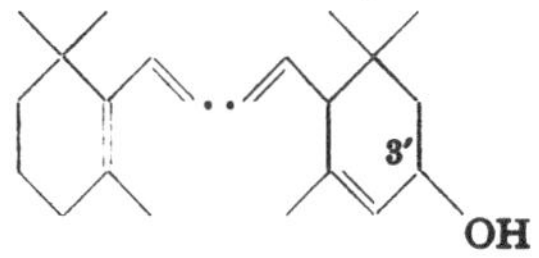

(XVIII.) α-Cryptoxanthin (3'-hydroxy-α-carotene).

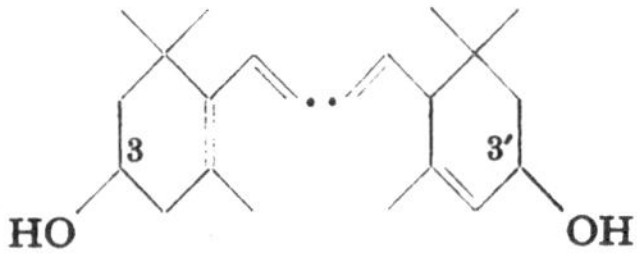

(XIX.) Lutein (3.3'-dihydroxy-α-carotene).

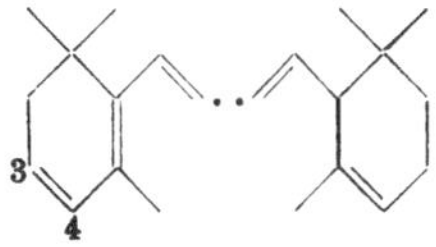

(XX.) 3,4-Dehydro-α-carotene.

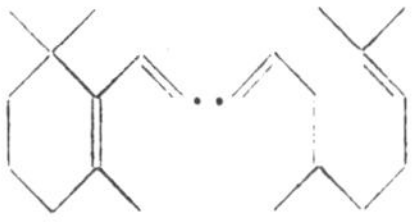

(XXI.) γ-Carotene.

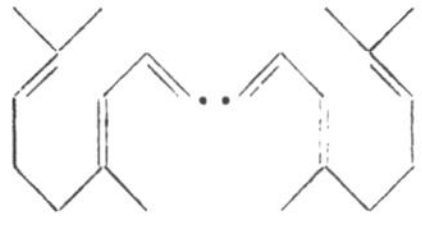

(XXII.) Lycopene.

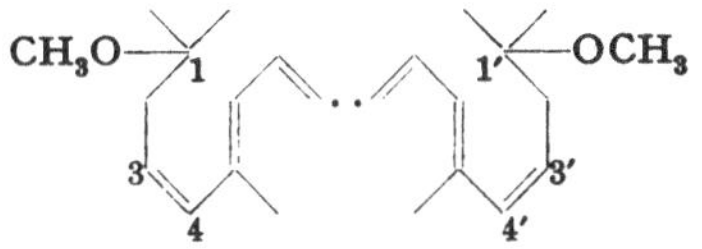

(XXIII.) Rhodoviolascin (spirilloxanthin) (23, 214).

A "stereoisomeric (or *cis-trans*-isomeric) equilibrium mixture" is present in a solution after a rearranging treatment has been applied to an all-*trans* compound or any other member of the set. A *cis* isomer which is encountered in substantial amounts is termed a "main" or "preferred" isomer.

Cis isomers with unknown configurations have been given the prefix "neo". In our laboratory neo A, neo B, etc. also mean that the *cis* isomer appears below the corresponding all-*trans* zone on the chromatographic column, while neo U (U for ultra), neo V, etc. indicate a location above the all-*trans* compound. Such differentiation does not always appear in older papers.

Some Historical Remarks on the Stereoisomerism of Polyenes.

In the following paragraphs only a few milestones in the history of *cis-trans* isomerism will be mentioned.

In 1819 a French pharmacist, POUTET (*371*), shook olive oil with a reagent that had been obtained by dissolving mercury in excess nitric acid and observed the solidification of the oil. POUTET was, of course, unaware of the fact that he had performed a *cis → trans* rearrangement of some unsaturated fatty acids about half a century before the foundation of modern stereochemistry was laid by VAN T'HOFF, LE BEL and others.

As is well known, VAN T'HOFF developed the basic, still valid theory of geometrical isomerism for ethylenic compounds (1875). The terms *"cis"* and *"trans"* were introduced by BAEYER much later (*18*).

Considerable time had elapsed before the successful stereochemical study of simple ethylenes could be extended to polyenes. The situation as it appeared in 1930 was correctly outlined by WITTIG and WIEMER (*484*) as follows:

> ,,Daß die *cis-trans* Isomerie mit steigender Zahl der konjugierten Doppelbindungen in den Hintergrund tritt und schließlich ganz verschwindet, obwohl die Zahl der Raumisomeren rasch zunehmen müßte, ist dann auf die wachsende Beweglichkeit der Valenzelektronen zurückzuführen, die die Isomerisationen zu den energieärmsten, stabilsten Lagen erleichtert. Ein lehrreiches Beispiel hierfür sind die von R. KUHN dargestellten Diphenyl-polyene $C_6H_5(CH=CH)_xC_6H_5$, deren Farbe sich mit wachsender Kettenlänge vertieft. Während die niederen Glieder der Reihe mit einer und mit zwei Doppelbindungen *cis-trans*-Isomere bilden, sind die höherkonjugierten Derivate nur in einer Raumform zu isolieren.''

This interpretation of the few then available data was, however, not accepted by KUHN and WINTERSTEIN (*270*):

> ,,Wir sind geneigt, die sterische Einheitlichkeit der höheren Diphenyl-polyene wenigstens teilweise den zur Synthese angewandten Methoden zuzuschreiben, ohne die Existenzfähigkeit entsprechender *cis*-Formen grundsätzlich zu bezweifeln.''

KUHN (*258*) has also drawn the attention to the natural product bixin as an example of *cis-trans* stereoisomeric polyenes (p. 107).

Subsequent developments have shown that long conjugated aliphatic systems undergo *trans → cis* rearrangements even more easily than do isolated *trans* double bonds.

The broad field of *cis-trans* isomeric C_{40}-carotenoids was opened up by GILLAM's pioneer studies in 1935–1936. GILLAM and EL RIDI (*129, 130*) observed that, after *repeated* adsorptions on alumina columns, β-carotene separated into two zones, an upper one containing unchanged starting material and a lower one, termed pseudo-α-carotene, a new product. According to GILLAM this isomerization process is reversible and leads to the equilibrium, β-carotene $\rightleftarrows \psi$-α-carotene. α-Carotene (*131*) behaves similarly and forms the equilibrium, α-carotene $\rightleftarrows$ neo-α-carotene.

These changes were originally attributed to an action of the adsorbent on the carotene: ". . . the analytical process itself affects the substances which it is designed only to separate." It was soon shown, however, by the writer in collaboration with TUZSON that the phenomenon is independent of the adsorption (*534, 535*). Indeed, it takes place spontaneously when a carotenoid solution is kept at room temperature for several hours or days. In the case of the tomato pigment lycopene, a continuous displacement of the spectral maxima towards shorter wavelengths can be followed in the visual spectroscope, i. e. without the use of an adsorbent (*536*).

It came to light that the lengthy operations, also including evaporation of solutions, were responsible for the heterogeneity discovered by GILLAM; this was confirmed by CARTER and GILLAM (55).

As expected, the reaction rates increased at higher temperature (refluxing). A further, particularly strong promoting effect was observed even at room temperature upon the addition of catalytic amounts of iodine, in light [writer and TUZSON (535, 536; cf. 367, 512)].

Independently, STRAIN (416) had noticed (1938) that several leaf xanthophylls were reversibly altered by heat or some other agents and yielded complex chromatograms. While characteristic spectral data were reported, the thermal effect and the alleged adsorbent effect were not yet clearly differentiated in these early studies (cf. 505).

As a theoretical interpretation of the isomerization GILLAM considered at first both geometrical isomerism and migration of a terminal double bond out of conjugation (130, 131). The latter assumption seemed reasonable at that time since, in contrast to β-carotene molecules which possess eleven conjugated double bonds, the color and spectrum of ψ-α-carotene corresponded to the presence of only ten such bonds.

When studying the behavior of the main red pepper pigment, the polyene-ketone capsanthin (ex Capsicum annuum), CHOLNOKY and the writer (502) reported in 1937 on some isomerization phenomena quite similar to those described by GILLAM. As possible explanations double bond migration, cis-trans rearrangement and keto-enol isomerism were considered.

When more experimental data had become available, it could be claimed with increasing emphasis that all conversions mentioned were trans → cis rearrangements (536, 504, 503). Valuable information was gained by studying the behavior of β-carotenone whose chromophore is blocked at both ends by conjugated carbonyl groups. Although this structure excludes double bond migration, the observed isomerization phenomena were the same as in the case of carotenoid hydrocarbons (504). Furthermore, as was pointed out later by HUNTER et al. (181), since the optical activity was maintained during the conversion, α-carotene → neo-α-carotene (131), no movement of the terminal 5,6-double bond could have taken place. The final decision in favor of geometrical isomerism was made on the basis of the high number of reversibly formed pigments, e. g. about a dozen in the case of β-carotene [POLGÁR and the writer (365)]. No other theory would allow for this richness of forms. Although this development encountered considerable scepticism at first [KARRER and RUTSCHMANN (235)], the basic concepts mentioned now appear to have been generally accepted.

A new epoch in polyene stereochemistry was initiated in 1950 when KARRER and EUGSTER (226) and, almost simultaneously, INHOFFEN et al. (*184*) reported the first total syntheses of β-carotene. In the last phase of these and several other syntheses certain *cis* intermediates appeared whose configurations followed beyond doubt from the very reaction sequence (cf. Chapter VI, p. 58).

In the course of the short history of polyene stereochemistry it has become increasingly clear that the carotenoid molecule is "morphologically sensitive" to spatial variations (stereomutation) in the sense that a single *trans* → *cis* shift may drastically alter the overall shape of the molecule. This feature is not shared by some other types of compounds which are rich in steric forms; thus, the *general* pattern of a sugar molecule is but little modified by epimerization.

Survey articles: CROMBIE (*73*), MACKINNEY (*296*), GOODWIN (*137*, *135*), WYMAN (*486*), BRODE (*43*), AMES (*9*), writer (*492*, *498*). For a bibliography of papers published by our research group see (*500*).

II. Number and Types of *cis* Carotenoids. Steric Hindrance.

As mentioned, the carotenoids and related multiconjugated systems constitute a unique case among low-molecular weight substances because of the great number of possible geometrical isomers.

In order to calculate the total number of possible *cis-trans* isomers (*516, 269*) the polyenes may be divided into "unsymmetrical" and "symmetrical" types; in the first type the two halves of the molecule are dissimilar but in the second type they are identical. If the conjugated system contains n aliphatic double bonds, the number of possible stereoisomers N will be:

For unsymmetrical systems: $N = 2^n$.
For symmetrical systems, n odd: $N = 2^{(n-1)/2} \cdot (2^{(n-1)/2} + 1)$.
For symmetrical systems, n even: $N = 2^{(n/2)-1} \cdot (2^{(n/2)} + 1)$.

Some pertinent values are listed in *Table 1*.

Table 1. Total Number N of *cis-trans* Isomers Calculated for Polyenes Containing n Sterically Effective Double Bonds.

n	N (symmetrical molecules)	N (unsymmetrical molecules)	n	N (symmetrical molecules)	N (unsymmetrical molecules)
1	2	2	7	72	128
2	3	4	8	136	256
3	6	8	9	272	512
4	10	16	10	528	1024
5	20	32	11	1056	2048
6	36	64	12	2080	4096

As will be shown in Chapter V (p. 46), any ordinary (all-*trans*) carotenoid can be converted by various treatments into a (quasi-) equilibrium mixture of *cis-trans* isomers. Theoretically, such a mixture should contain all possible spatial forms. Furthermore, when isolated from the mixture and submitted again to rearrangement, each stereoisomer should yield the same equilibrium. This postulate has been verified experimentally. Indeed, it must be fulfilled before an observed pigment can be accepted as a member of a given stereoisomeric set.

At present, there still exists a wide gap between the number of calculated and observed spatial forms of carotenoids. Experiment has shown that each carotenoid yields only a very limited number of *cis* isomers, mostly from two to a dozen. In no case did more than two or three main *cis* pigments appear whose respective quantities amounted to at least 10% of the stereoisomeric mixture.

Although the differences between the thermodynamic behavior of the many possible isomers do not seem to be excessive, the action of a selective mechanism is manifest whose nature is but partly understood. Contributing factors seem to be the relative activation energies of the respective all-*trans* → *cis* rearrangements and the distribution of the electronic charges in the individual isomers. These factors, rather than the relative stability of a given *cis* form, may determine the importance of the latter in the set (cf. *376, 32, 375*).

It is hoped that the configurations of an increasing number of *cis* isomers will be clarified in the future. Then is should become possible to work out a theory concerning the relationship between a given configuration, its relative stability and the probability of its appearance in a *cis-trans* isomeric equilibrium. At present the following points (a)–(c) can be made as an attempt to narrow the gap between the number of calculated and observed spatial forms of carotenoids.

(a) The composition of equilibrium mixtures is influenced by the chemical structure in the sense that it will be more complex in a non-symmetrical than in an analogous symmetrical set. In the latter instance each double bond (except the central one) finds its stereochemically equal counterpart in the other half of the molecule and thus the total number of possible *cis* configurations is reduced. The probability of formation of a given preferred *cis* double bond increases by a factor of two. The observed composition of the *cis-trans* isomeric mixture is simpler in some symmetrical than in comparable unsymmetrical sets.

Examples (*525, 365*). The relative amounts of the *cis* isomers were found as follows: α-Carotene set (unsymmetrical), neo U : neo V : neo W : neo B : minor *cis* isomers = 15 : 3 : 16 : 13 : 3; β-carotene set (symmetrical), neo U : neo B : minor isomers = 22 : 25 : 5. The unsymmetrical all-*trans*-γ-carotene yields a very complex equilibrium mixture but in the case of the symmetrical lycopene practically only one main *cis* isomer (neo A) is present, to the extent of 50% of the total pigment.

(b) A second restricting factor is the spectroscopically established feature that the rearrangement of all-*trans* carotenoids yields mainly mono*cis* and di*cis* (but not poly*cis*) forms (cf. λ_{max} shifts, pp. 28, 78). This is understandable as follows. Suppose that the equilibrium ratio, mono-*cis*/all-*trans* is of the order of magnitude $^{1}/_{10}$. If it is assumed for the sake of simplicity that roughly the same change in free energy accompanies a *trans* → *cis* rotation about each double bond, then this ratio, e. g. for a tetra*cis* form, would amount to only $^{1}/_{10\,000}$. In other words, the isomer would not be detectable by current analytical methods (*515*).

That indeed the probability of the appearance of a given configuration decreases rapidly with the increasing number of *cis* double bonds, was demonstrated by PETRACEK and the writer (*519*) in the following, unusually large-scale experiment; it has shown that the order of magnitude of poly*cis* lycopenes in thermally obtained equilibria must be less than 1 part in 3 million parts of the total pigment.

A benzene solution of 30 g. of pure, crystalline all-*trans*-lycopene was refluxed for 30 min. and submitted to careful chromatographic fractionation. None of the pigment fractions showed upon iodine catalysis the migration of the spectral maxima typical for poly*cis* forms (Fig. 18, p. 68). This test turned, positive, however, when 40 µg. of the poly*cis* compound prolycopene was introduced. Further, as little as 10 µg. of added poly*cis* lycopene when washed through the chromatographic column was detected easily in the first liter of the filtrate.

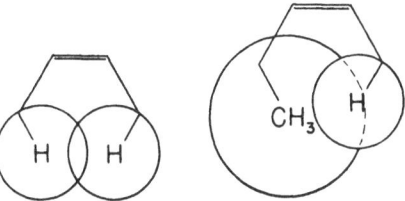

Fig. 1. Overlapping of hydrogen atoms in —CH—CH=CH—CH— and of hydrogen and methyl in —CH—CH=CH—CCH$_3$ with *cis* configuration; according to PAULING (*347*). [From: Fortschr. Chem. organ. Naturstoffe 3, 203 (1939).]

(c) In order to evaluate a third and important restricting factor, the double bonds of a carotenoid are divided into two types, termed, respectively, "unhindered" and "hindered" ones. This distinction was proposed by PAULING (*347*):

A *cis* double bond located in a polyene chain will have the orientation,

$$\begin{array}{ccc} \text{H} & & \text{H} \\ \diagdown & & \diagup \\ & \text{C}=\text{C} & \\ \diagup & & \diagdown \\ =\text{C} & & \text{C}= \\ \diagdown & & \diagup \\ & \text{X} \quad \text{X}' & \end{array}$$

In carotenoid molecules X and X' represent either hydrogen atoms or methyl groups. When both X and X' are hydrogens the formation of the *cis* isomer by rotation about a *trans* double bond does not encounter any significant steric hindrance. This, however, does happen when X represents a hydrogen and X' a methyl *(Fig. 1)*. Then a *trans* → *cis* rearrangement is opposed by a considerable strain, and a significant energy barrier must be overcome.

Inspection of the β-carotene formula (I, p. 5), for example, will show that the molecule contains both types of double bonds, viz. out

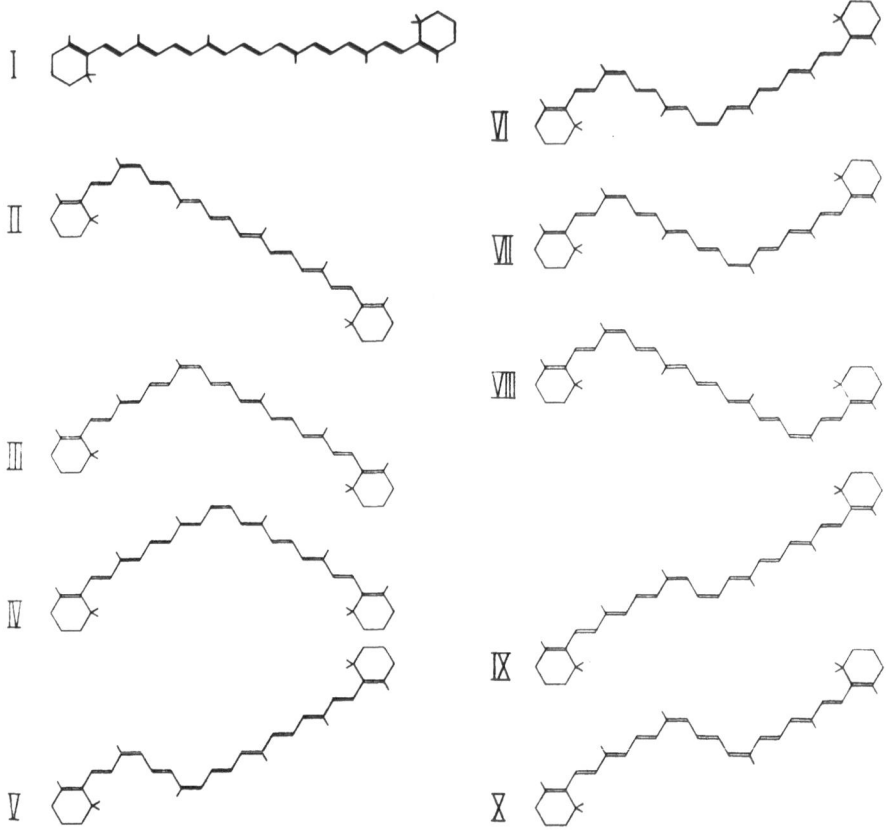

Fig. 2. Skeleton models of the twenty possible unhindered stereoisomers of β-carotene: all-*trans*-β-carotene, three mono*cis*-β-carotenes, six di*cis*-β carotenes, six tri*cis*-β-carotenes, three tetra*cis*-β-carotenes, and one penta*cis*-β-carotene *(492)*. [From: Chem. Rev. **34**, 267 (1944).]

of the nine aliphatic such bonds only five are unhindered, namely the four that carry methyl groups (positions 9, 13, 13′, 9′) and the central double bond (position 15).

When an all-*trans* carotenoid is subjected to stereoisomerization, the formation of unhindered and hindered *cis* isomers will compete with each other and the former type will be given preference.

Recent work has shown that the presence of a hindered *cis* double bond in a polyene molecule degrades the spectral curve in the visible region in the sense that the extinction values and the degree of fine structure decrease substantially; thus, the fundamental band flattens (cf. Figs. 27–28, p. 83). Degraded spectra did, however, as a rule not appear on *trans* → *cis* rearrangements and hence no hindered *cis* carotenoids were formed. *Table 2* (p. 16) shows the practical importance of this simplification: the calculated

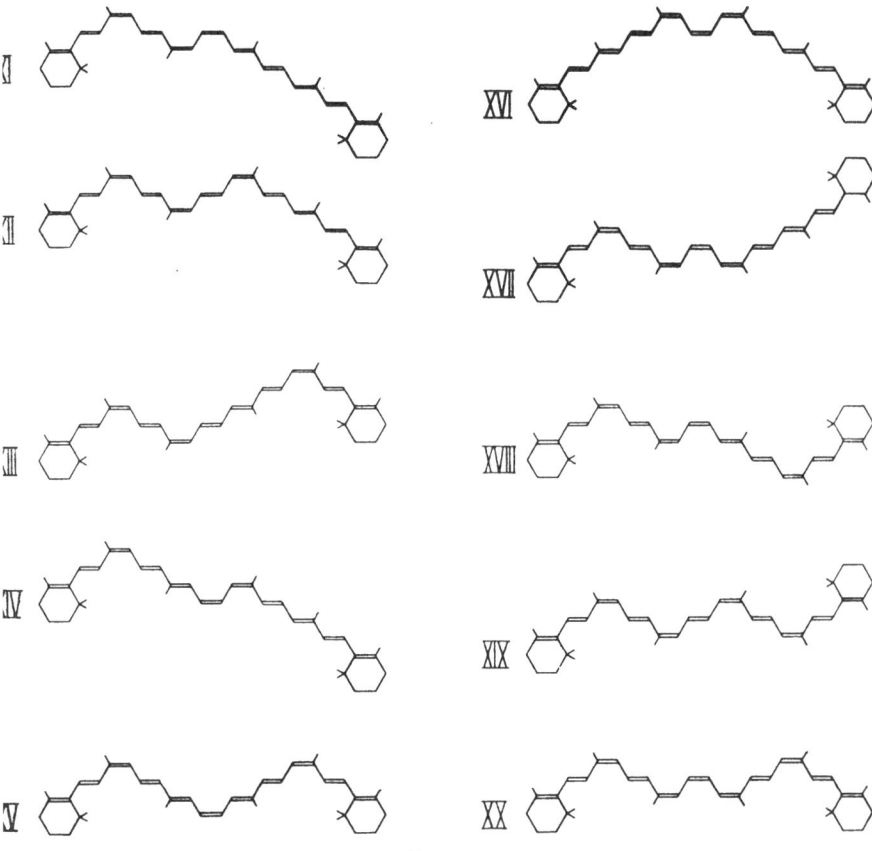

Fig. 2.

number of unhindered *cis* isomers amounts to only 6–7% of that of the total possible isomers.

In the case of β-carotene only 20 unhindered isomers were expected and 12 were observed (*365*)—a reasonably good agreement. *Fig. 2* represents the skeleton models of the 20 unhindered β-carotenes.

To summarize, several factors have been helpful in the interpretation of the existing gap between the calculated and found number of isomers, viz. structural features, the low probability of the formation of poly*cis* forms, and the sterically hindered nature of some double bonds.

In exceptional instances small amounts of hindered *cis* carotenoids were obtained by melting crystals (p. 47).

The restriction to unhindered isomers in direct rearrangement experiments is invalid in the vitamin A, retinene and some unbranched diphenylpolyene sets (pp. 125, 149).

Table 2. Calculated Number of the Sterically Unhindered and Total
cis-trans Isomeric Forms of Certain C_{40}-Carotenoids.

Name	Symmetrical (s) or unsymmetrical (u) structure	Number of aliphatic, conjugated C=C double bonds	Number of theoretically possible steric forms	
			unhindered	total
α-Carotene	u	9	32	512
β-Carotene	s	9	20	272
γ-Carotene	u	10	64	1024
Lycopene	s	11	72	1056
Lycoxanthin	u	11	128	2048
Cryptoxanthin	u	9	32	512
Zeaxanthin.	s	9	20	272
Lutein	u	9	32	512
Astacin	s	9	20	272
Capsanthin	u	9	32	512
β-Carotene mono-epoxide	u	9	32	512
Violaxanthin (zea-xanthin diepoxide)	s	9	20	272

Two decades ago, when PAULING (*347*) developed his theory of hindered and unhindered double bonds, the only available way for the preparation of *cis* carotenoids was direct rearrangement. Subsequently, several authors (among them the present writer) more or less tacitly interpreted his "rule" as being prohibitive for the existence of hindered *cis* forms in the C_{40}-carotenoid series. It can now be stated as the outcome of a controversy (*237, 349, 102, 106, 125, 128*) that while it is indeed difficult or impossible to overcome the energy barrier in the direct process, *trans* → hindered *cis*, this obstacle can be circumvented by following an entirely different route, namely total synthesis. As will be shown in Chapter VI (p. 58) a triple bond is built up at the site of the prospective hindered *cis* double bond and is then catalytically semi-hydrogenated in a stereospecific manner to give a *cis* olefin. Thus, a hindered *cis* carotenoid can be obtained in good yield.

The first clear exception from the "rule" was reported by OROSHNIK, KARMAS and MEBANE (*337*) in the *retro*-vitamin A methylether set. Soon after the same research group synthesized hindered *cis* isomers of several isoprenic dienes, trienes, tetraenes and pentaenes of the C_{20}-series (*339*). KARRER et al. were successful in synthesizing hindered *cis* forms of β-carotene, lycopene, and 1,18-diphenyl-3,7,12,16-tetramethyl-octadecanonaene (p. 104) (*106, 125, 126, 128, 287*). Even such hindered *cis* isomers were prepared that carried phenyl, instead of methyl, side-chains (*540*) (p. 105).

A slightly different type of steric conflict would be caused by a *s-cis* conformation at the 6,7-bond, viz. interference of the 8-hydrogen atom with the 5-methyl group. For 15,15'-dehydro-β-carotene crystals the *s-trans* conformation has been proven by SLY's X-ray studies (*407*), and there is no reason why this finding should not be valid for solutions of the common carotenoids.

As has been pointed out by DALE (*79*), the *s-trans* conformation causes an interference between the *gem.*-dimethyl group and the opposing hydrogen atom also in all-*trans*-carotenoids and analogous compounds that possess β-ionylidene or dehydro-β-ionylidene structures. Consequently, for example, β-carotene and the vitamins A show only a moderate degree of fine structure and lower than expected extinction values in the main band.

s-cis (6,7)
(interference of the 8-hydrogen with the 5-methyl).

s-trans (6,7)
(interference of the 8-hydrogen with one of the 1-methyls).

Further literature, OROSHNIK and MEBANE (*339*), WAIGHT and ERSKINE (*447*), BRAUDE and WAIGHT (*41*).

III. Some Properties of *cis* Carotenoids.

1. Relative Stabilities.

An all-*trans* carotenoid, because of its nearly coplanar chromophore and perfect resonance, is generally expected to surpass in stability its *cis* isomers. Nevertheless it should be pointed out that simple characterizations such as "stable" or "labile" have lost much of their meaning in the class of multiconjugated systems. Because the degree of the double bond character is subject to considerable variation along the chain, the extent of lability of a given steric form depends not only on the number but also on the location of *cis* double bonds, hence on the *overall* shape of the molecule.

Sterically Unhindered cis Forms. Various stereoisomerizing treatments (especially thermal as compared to photochemical) may affect individual configurations quite differently. For instance, during the (theoretical) transition, all-*trans* → mono*cis* → di*cis* → poly*cis*, the thermostability in solution first decreases and then increases again when the bent molecule gradually straightens out and reassumes a compact overall shape. This behavior is contrasted with the increase of the photochemical sensitivity in the presence of iodine during the entire process.

Considerable variation in the stereostability has also been noted during crystallization. All-*trans* pigments and certain natural and artificial *cis* forms crystallize easily and without change. In other instances, however, either no crystallization takes place or spatial rearrangement competes with the formation of the crystal grating. Thus, neo-β-carotene B (GILLAM's ψ-α-carotene, p. 8) crystallizes easily but (in our hands) even fresh solutions of the crystals contained some all-*trans* form and afforded complex chromatograms. Still more sensitive is the configuration of the main lycopene isomer, neo A, that yields exclusively all-*trans* crystals (*535*). The closely related, synthetic central-mono*cis*-lycopene rearranges to the all-*trans* pigment when heated for a short time during recrystallization (*197*).

Recent studies into the behavior of several central-mono*cis* carotenoids have shown that this highly bent molecular shape causes strong photosensitivity (p. 77).

Several mono*cis* carotenoids of the neo U type which according to the spectra (cf. p. 76) must possess a peripherally located *cis* double bond and a slightly bent molecular shape, are distinguished by a high degree of thermostability. They remain unchanged when the solution is refluxed and crystallize rapidly on cooling.

Examples. Neo-α-carotene U and neo-β-carotene U (*525, 365*).

Sterically Hindered cis Forms. As we have seen, such isomers, once formed, show considerable photo- and thermostability. Some of them are well crystallized and can be heated almost to their melting points without suffering rearrangement.

Examples. 11,11'-Di*cis*-β-carotene (*cis*-β-carotene B) (*107, 195*); *cis*-β-carotene C (*107*); 5-*cis*-1,18-diphenyl-3,7,12,16-tetramethyl-octadecanonaene and its 5,13-di*cis* isomer (*125*); several *cis* 1,3,7,12,16,18-hexaphenyl-octadecanonaenes (*540*). The hindered *cis* lycopenes b, c and c′ are much more labile and could not be crystallized (*128*).

All known sterically hindered *cis* carotenoids are extremely sensitive to iodine, in light, and the stereochemical equilibrium is reached at much higher rates than in the case of an all-*trans* or unhindered *cis* isomer.

Polycis Forms. The behavior of the naturally occurring poly*cis* carotenoids will be discussed in Chapter VII, p. 63.

2. Melting Points.

The situation regarding the solubilities and melting points is simple, since in each stereochemical set the least soluble and highest melting member is the all-*trans* compound. The melting point depression caused by a *trans* → *cis* rotation is considerable and amounts to 20–120°. The effect is not proportional to the number of *cis* double bonds. Neither hindered nor poly*cis* isomers show any particular behavior in this respect (*Table 3*, next page).

In some instances unsharp melting points are observed because of partial rearrangement in the fused or sintered state; some isomers are converted rapidly into the corresponding all-*trans* compounds and show the phenomenon of double melting points.

Examples. (a) Central-mono*cis*-lycopene melts at about 105°, then solidifies and melts again at higher temperature (*197*). (b) Central-mono*cis*-3,4,3′,4′-bis-dehydro-β-carotene melts between 135° and 140°, and for the second time at 193° (m. p. of the all-*trans* form, 196–197°) (*191, 204*). (c) 5,13-Di*cis*-1,18-diphenyl-3,7,12,16-tetramethyl-octadecanonaene, a hindered isomer, darkens considerably at 155° and melts at 210° which is the m. p. of the all-*trans* form (*106*).

Table 3. Examples of Melting Point Depression Caused by *trans* → *cis* Isomerization of Carotenoids.

Stereoisomeric set	M. p. of all-*trans* form (°)	M. p. of *cis* isomer (°)		References
α-Carotene	187–188	Neo U	65	(525)
β-Carotene	183	Neo U	122–123	(365)
		Neo B	146	(397)
		Central-mono*cis*	151	(185, 187, 203)
		11,11'-Di*cis* (hindered)	154	(107, 195)
3,4-Dehydro-β-carotene	186	Central-mono*cis*	138	(204)
3,4,3',4'-Bisdehydro- β-carotene	196–197	Central-mono*cis*	135–140	(204)
16,16'-*Homo*-β-carotene $C_{42}H_{58}$	190	Middle-di*cis*	142	(183)
13,13'-Bis-desmethyl- β-carotene $C_{38}H_{52}$	177	Central-mono*cis*	157	(186)
γ-Carotene	178; 150	Neo P	89–90	(521, 529)
		Pro-γ-carotene	134–135	(531)
Lycopene	175	Central-mono*cis*	105	(197)
		Prolycopene	111	(282)
		Poly*cis* I	93–95	(521)
		Poly*cis* II	85–87	(521)
		Poly*cis* III	105–106	(521)
retro-Dehydro-carotene	192–193	Neo A	161	(537)
		Neo D	177–178	(537)
Cryptoxanthin	169	Neo A	75–76	(59)
		Neo U	95–96	(59)
Zeaxanthin	205	Neo A	106	(504)
		Neo B	92	(504)
		Neo C	154	(504)
Canthaxanthin	213	Central-mono*cis*	161–162	(123)
		Neo A	136–137	(206)
		Neo B	181	(206)
β-Carotene diepoxide	189–190	Neo V	177–178	(440)
Antheraxanthin (zea- xanthin monoepoxide)	205	Mono*cis* (natural)	110	(432)

3. Rotatory Power.

In the presence of an asymmetric carbon atom, *trans* → *cis* rotations involve changes in the optical activity.

Thus, according to STRAIN's early observations (416) and those made by our group (504), the *levo* rotation of natural zeaxanthin is inverted to *dextro* by heating or by iodine catalysis. The resulting solution contains a strongly dextrorotatory isomer. It was also found (532) that all-*trans*-gazaniaxanthin, whose rotatory power was unmeasurably small, afforded with iodine a stereoisomeric mixture showing $[\alpha]_D^{20} = +160°$. After chromatography, some weakly adsorbed *cis* forms had the value $+ 220°$ that decreased to $+ 155°$ when the catalytic treatment was repeated.

Such experiments should be evaluated with caution, especially when the rotatory power decreases, since racemization and/or irreversible cleavage may interfere. It should be demonstrated in every case that, upon re-isomerization of a *cis* form an all-*trans* sample is obtained that shows the initial optical activity.

Table 4. Examples of the Influence of the Configuration on the Rotatory Power.

Stereoisomeric set	Solvent	Steric form	$[\alpha]_{Cd}^{25}$ or $[\alpha]_{C}^{20}$	References
α-Carotene	chloroform	All-*trans*	$+ 359°$	(525)
		Neo U	$+ 221°$	
Gazaniaxanthin	petroleum ether	All-*trans*	$\pm 0°$	(532)
		Neo-Group I	$+ 220°$	
Zeaxanthin	chloroform	All-*trans*	$- 42.5°$	(504)
		Neo A	$+ 120°$	
		Neo B	$\pm 0°$	
Capsanthin	benzene	All-*trans*	$\pm 0°$	(503)
		Neo A	$+ 89°$	
		Neo B	$+ 21°$	
		Neo C	$+ 27°$	
Capsanthin dipalmitate	petroleum ether	All-*trans*	$- 30°$	(503)
		Neo A	$- 22°$	
		Neo B	$- 20°$	
Capsorubin	benzene	All-*trans*	$\pm 0°$	(503)
		Neo A	$- 134°$	
		Neo B	$- 69°$	
Capsorubin dipalmitate	petroleum ether	All-*trans*	$\pm 0°$	(503)
		Neo A	$- 75°$	
		Neo B	$- 15°$	

Table 4 refers to crystalline products. We note that in some instances the rotatory power of one steric form by far surpasses that of all others. On the basis of spectral readings such isomers seem to be mono*cis* compounds containing the *cis* double bond at or near the center. It is still unknown which particular configurational type is responsible in general for a substantial increase of the described manifestation of the molecular asymmetry. Both theoretical and experimental problems would merit closer attention in this interesting field.

4. Relative Adsorption Affinities.

It has been established empirically, in the absence of a pertinent theory, that the relative adsorption affinities of stereoisomeric carotenoids are so sharply dependent on the configuration that the resolution of even complicated mixtures can be realized easily. Indeed, the systematic use of chromatography has made possible substantial progress in the polyene field. In considerable contrast, classical methods such as fractional crystallization may be satisfactory in the field of simple olefins.

Column chromatographic techniques have been applied with advantage in countless instances to the resolution of *cis-trans* isomeric mixtures and isolation of stereochemically pure, individual *cis* carotenoids. While the reader is referred to the monographs (*493, 494, 492, 501, 497, 419, 278, 232*), only a few practical hints can be offered at this point.

Frequently used adsorbents are, calcium hydroxide (+ Celite), calcium carbonate, alumina, magnesia, zinc carbonate; and developers, hexane, petroleum ether, benzene, hexane-benzene mixtures, and hexane containing a polar solvent (*497*).

In our laboratory, a series of hexane-acetone mixtures are kept ready for use, ranging from 99.5 : 0.5 to 80 : 20. When a new substance is to be studied, first pure hexane, then hexane with a low acetone content is tried out as a developer. The acetone content is increased gradually until optimum resolution takes place.

At the end of the developing process we extrude the column and excise the individual zones; fractional washings are often less reliable. One can, however, allow weakly adsorbed components such as carotenes and phytofluene, to pass into the filtrate, and separate by cutting the more strongly adsorbed pigments. In order to detect fluorescent impurities, inspection of the chromatogram by means of a portable ultraviolet lamp, under a black cloth, is recommended.

Each individual pigment is eluted with a solvent more polar than the developer such as hexane-acetone 1 : 1. After addition of water, the acetone can be removed by using LE ROSEN's automatic device (*280*).

For relative eluting strengths of solvents and R_f values of *cis* β-carotenes, see BICKOFF (*33*). — Combination of column chromatography and countercurrent distribution: CURL (*75–77*). — Paper chromatography of carotenoids: GRANGAUD and GARCIA (*143*), BAUER (*29*), LEDERER and LEDERER (*278*), JENSEN and JENSEN (*213*). This method has been little used for separating stereoisomeric polyenes.

Tables 9 and 11 (pp. 48, 53) list a few data concerning relative positions of stereoisomeric carotenoids on the column.

Unfortunately, theory is lagging behind practical experimentation in this field, perhaps because very little is known about the orientation of individual carotenoid isomers on an active surface or about the strength of the adsorption forces as a function of the molecular form. Nobody has yet been able either to predict or to explain why a *trans* → *cis* rotation, i. e. the bending of a rod-like, all-*trans* carotenoid molecule may either increase or decrease the adsorption affinity with the result that in a chromatogram certain *cis* zones appear above but others below the unchanged portion of the all-*trans* pigment.

Evidently, this dual effect has a character quite different from the all-*trans* → *cis* spectral shift which invariably weakens the color of carotenoids. Under the influence of spatial rearrangements adsorbability and light extinction may change either in a parallel or in an opposite sense.

Let us endeavor now to summarize some aspects of the relationship between adsorption behavior and geometrical isomerism. When evaluating the following paragraphs it should be kept in mind that any statement on adsorption forces is valid only for a given system, consisting of substance, adsorbent, and developer. By changing the adsorbent or developer or both, the resulting chromatogram may be altered substantially.

We assume that not every section of an adsorbed organic molecule will be equally responsible for the fixation process but certain "anchoring groups" will play the decisive part (*494, 497, 493*). In favorable instances the anchoring group can be identified experimentally by demonstrating that modification or elimination of that particular section of the molecule causes an unusually drastic change in the adsorption behavior (cf. some statements on dihydroxy-carotenes and their esters on p. 24). Within a complex molecule two or more potential anchoring groups may compete for the active spots of the adsorbent, and it seems that in principle more than a single orientation of the molecule is possible on the surface. Consequently, certain changes in the chromatographic system may induce even an inversion of the top-to-bottom sequence of two substances.

As a characteristic example, the behavior of the pair, lycopene (IV, no rings, 11 conjugated and 2 isolated double bonds, no OH-group) and cryptoxanthin (VIII, 2 rings, 11 conjugated double bonds, 1 OH-group) (pp. 6–7) may be mentioned. On benzene-developed alumina or calcium carbonate columns the sequence is, cryptoxanthin (top) and lycopene (bot-tom) which indicates that the hydroxyl group is responsible for the stronger fixation of cryptoxanthin; its effect overrules that of the greater number of double bonds present in lycopene. The top-to-bottom sequence is inverted, however, on lime where the hydroxyl does not seem to function as the (main) anchoring group (*281*). Some other impressive inversions were reported by STRAIN (*420–422*). It would be well worth while to investigate how far inversion phenomena can be realized pertaining to the chromatographic sequence of several *cis* forms of a given carotenoid.

Let us now compare the changes in adsorption affinities that are induced, respectively, by stereochemical and by (reasonably chosen) structural alterations of the molecule. Experiments have shown that structural and stereochemical factors are here of equal order of magnitude.

To illustrate, it was rather surprising to observe that the difference between the adsorption affinities of all-*trans*-β-carotene (VII, p. 6, 11 conjugated double bonds; top position) and all-*trans*-α-carotene (XVII, p. 7, 10 such bonds; bottom position) can be overruled by bending suitably the α-carotene molecule. After a mixture of the two all-*trans* carotenes had been subjected to a treatment with iodine and subsequent chromatography, one of the *cis* α-carotenes (neo U) appeared *above* the all-*trans*-β-carotene zone (*525*).

The top-to-bottom sequence was: neo-β V, neo-α U, all-*trans*-β, neo-α V, neo-β B, neo-β E, neo-α W, neo-β F, all-*trans*-α, and neo-α B.

Were the effect of the structural difference between β- and α-carotene of a higher order of magnitude than that of configurational differences, then all observed members of the β-carotene set would be expected to appear in a top section of the column followed by the well-separated family of stereoisomeric α-carotenes.

Although the adsorption forces are profoundly influenced by the number of *cis* double bonds in a polyene molecule, the dependence is not a simple one, because the effect is also a function of the position of these bonds.

In the subclass of the carotenoid hydrocarbons, the entirely aliphatic lycopene yields exclusively such *cis* isomers that show weakened adsorption affinities and appear below the all-*trans* zone (*535, 536, 515*). A treatment of either of the three hydroaromatic carotenes, α-, β-, or γ-, results, however, in the formation of isomers some of which preceed and others follow the all-*trans* zone in the chromatogram (*365, 525, 526*). As we will see later in connection with some spectral phenomena, a *trans* → *cis* rotation in the middle section of the carotene molecule decreases the adsorption affinity, and this effect is especially strong in the case of a central-mono*cis* carotene. In contrast, the adsorption affinity of a peripheral mono*cis* carotene may surpass that of the all-*trans* form.

Upon the introduction of one hydroxyl group the described chromatographic behavior of α- or β-carotene remains essentially unaltered in the sense that some *cis* forms of, e. g., 3-hydroxy-β-carotene (VIII, p. 6) appear above but others below the all-*trans* zone (*536, 512, 59, 49*). An impressive change takes place, however, when a second OH-group is introduced: each *cis* zone is then found above the all-*trans* pigment. The best known representatives of this type are lutein (3,3'-dihydroxy-α-carotene) (XIX) and zeaxanthin (3,3'-dihydroxy-β-carotene) (X) (*416, 536, 504*). Possibly, the reduced distance between the two OH-groups in the bent *cis* molecules is responsible for this behavior; and perhaps both hydroxyls are able to participate in the anchoring process. Upon esterification, the "excess" adsorption affinity disappears and each *cis* isomer is retained below the all-*trans* zone.

Examples. Physaliene = zeaxanthin dipalmitate, and esters of the hydroxy-ketones capsanthin and capsorubin (*504, 503*). It may be mentioned that canthaxanthin (4,4'-diketo-β-carotene) yields *cis* isomers with weakened adsorption affinities (*123*). The *cis* β-carotene epoxides are adsorbed in part below and in part above the all-*trans* zone (*440*).

Much more uniform is the chromatographic behavior in the presence of several *cis* double bonds: each known poly*cis* carotenoid is adsorbed far below the corresponding all-*trans*, mono*cis* and di*cis* forms. Inversely, when a poly*cis* compound is submitted to *cis* → *trans* rearrangement under mild conditions, the stepwise formed isomers (with rare exceptions) display stronger adsorbabilities than does the starting material (*301*, cf. p. 70).

IV. *Cis-trans* Isomerism and UV Spectra.

1. Some Remarks on the Spectra of all-*trans* Carotenoids.

The intense color of carotenoids is caused by the strong extinction in the region, 400–500 mμ, where the spectral curve shows a massive band. If the latter exhibits vibrational fine structure, in most instances three maxima or two maxima and a shoulder appear, whereby the middle peak has the highest intensity.

As is well known, the spectral properties of an all-*trans* polyene are determined, first of all, by the number of conjugated double bonds. A second factor is the shape of the carbon chain that carries the chromophore. This is well illustrated by the considerable difference between the lycopene and the β-carotene spectra (Figs. 6 and 4, pp. 29, 27). Both pigments possess eleven conjugated double bonds; however, the lycopene chromophore is entirely aliphatic while the two ends of the β-carotene chromophore reach into non-planar cyclohexene rings. The transition, β-carotene → lycopene, has a strong bathochromic effect.

The presence in carotenoid chromophores of four characteristically located methyl side-chains also influences the spectrum. Thus, the λ_{max} of INHOFFEN's synthetic lower β-carotene homolog, 13,13'-bis-desmethyl-β-carotene (*186*) from which the two middle CH_3-groups are missing, is located at 10 mμ shorter waves than that of β-carotene. In contrast, the introduction of 2- and 2'-methyl groups displaces the β-carotene maxima only slightly, because the chromophore is not affected (*186*).

Interesting is the marked spectral difference between YAMAGUCHI's renieratene and isorenieratene (*487–489*); these natural products differ only with respect to the position of their aromatically bound methyl groups that interfere sterically with the main aliphatic chain (p. 105).

Inspection of the extinction curves of numerous all-*trans* carotenoids has shown that the fundamental band may or may not possess vibrational fine structure; the latter is especially sharp in the case of *retro* carotenoids, however, the individual differences are also considerable in the *normal* series. Aliphatic or semi-aliphatic pigments such as lycopene or γ-carotene show much more pronounced fine structure in the main band than do the hydroaromatic α- and β-carotenes. On lengthening one side of the β-carotene chromophore by a conjugated carbonyl group or a conjugated ring double bond the fine structure is reduced considerably and may disappear altogether when both sides of the system are affected.

Examples. 4-Keto-β-carotene, 4,4′-diketo-β-carotene; 3,4-dehydro-β-carotene, and 3,4,3′,4′-bisdehydro-β-carotene *(351, 221, 499, 191, 204, 520)*. The curves of the dehydro compounds mentioned are in accordance with the loss of (the slight) fine structure during the transition, vitamin $A_1 \rightarrow$ 3,4-dehydrovitamin A_1 (vitamin A_2).

For theoretical interpretations of vibrational fine structures in polyenes see PLATT *(363)*, DALE *(79, 80, 82)*, and H. KUHN *(255)*.

2. Spectral Effect of *trans* → *cis* Isomerization in the Visible Region.

The profound influence of the molecular shape on the selective absorption of light has secured a preponderant role to spectroscopic readings in the study of *cis-trans* isomeric polyenes.

It was found that when an all-*trans* carotenoid is converted into a mixture of *cis-trans* isomers (p. 46) the color intensity decreases. At relatively high pigment concentrations the effect may be followed by the naked eye and at low concentrations in the visual spectroscope.

To observe such phenomena in the visual spectroscope, first the position of the maxima of the all-*trans* compound is determined and then a drop of dilute

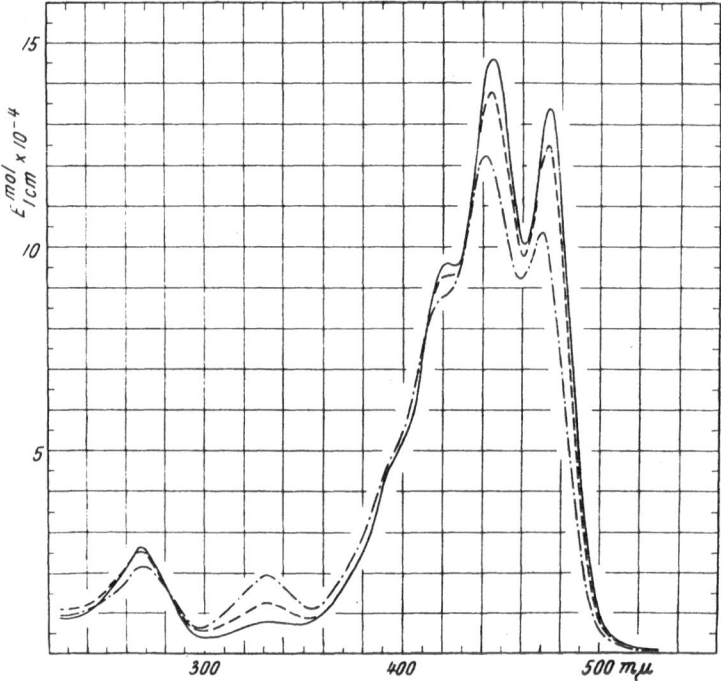

Fig. 3. Molecular extinction curves of α-carotene, in hexane: ————, fresh solution of the all-*trans* compound: — — —, mixture of stereoisomers after refluxing; and — · — · —, after catalysis by iodine *(524)*. [From: J. Amer. Chem. Soc. **65**, 1522 (1943).]

iodine solution is introduced (the light required for the catalysis is furnished by the apparatus itself). The bands become moderately blurred and migrate towards shorter wavelengths in 1–2 minutes; the new positions are then determined. When starting from a *cis* pigment, migration of the bands in the opposite direction takes place. Thus, the observer is able to differentiate, within minutes, between an all-*trans* and a *cis* carotenoid. The final wavelength position of the bands may also indicate the stereochemical set to which a *cis* isomer belongs.

It is recommended to characterize a polyene by recording its maxima both before and after catalysis by iodine, even when no further stereo-chemical work is planned. The same statement is valid for extinction curves.

During an all-*trans* → *cis* rearrangement the spectral curve is altered as follows in the visible region: The extinction values and the degree of fine structure decrease while the maxima migrate towards shorter wavelengths *(Figs. 3–6)*. Since the resulting equilibrium mixture usually contains 40–60% unchanged all-*trans* carotenoid, the spectral difference between the latter and the main *cis* isomer(s) evidently surpasses the observed shift. This can be demonstrated in somewhat larger scale

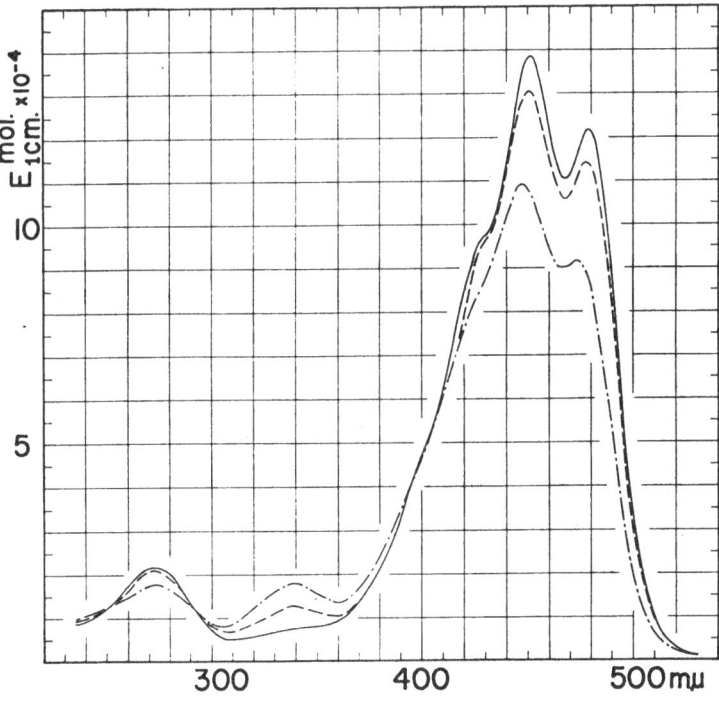

Fig. 4. Molecular extinction curves of β-carotene, in hexane: ————, fresh solution of the all-*trans* compound; — — —, mixture of stereoisomers after refluxing; and — · — · —, after catalysis by iodine (*524*). [From: J. Amer. Chem. Soc. **65**, 1522 (1943).]

experiments by chromatographic resolution of the equilibrium mixture and determination of the spectral curve of each individual *cis* isomer. Such curves undergo the following changes upon iodine catalysis: The extinction values at the maxima as well as the degree of fine structure increase and the maxima migrate towards longer wavelengths. Finally,

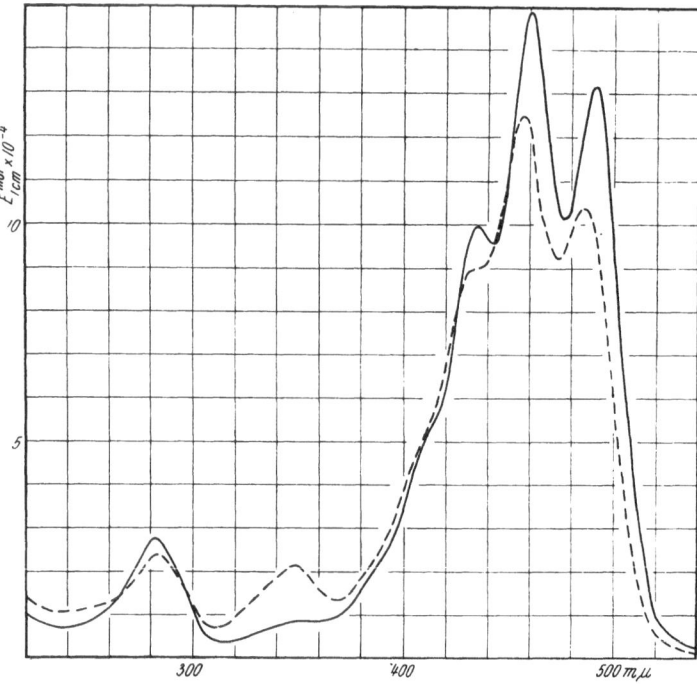

Fig. 5. Molecular extinction curves of γ-carotene, in hexane: ————, fresh solution of the all-*trans* compound; and — — —, mixture of stereoisomers after catalysis by iodine (*353*).

the spectrum of the same equilibrium mixture is obtained as on interaction of the all-*trans* compound and iodine.

The wavelength difference, in mμ, between the location of λ_{max} of an all-*trans* compound and that of one of its *cis* isomers is termed the "λ_{max} *shift*".

The validity of the statement that in the visible region an all-*trans* carotenoid shows longer wavelength maxima than any of its *cis* isomers can best be illustrated in the lycopene set of which about 40 spectroscopically well-defined members are known [Table 15, p. 73; (*301*)]. Some of them were prepared by rearranging the all-*trans* pigment, others by partial *cis* → *trans* isomerization of a poly*cis* lycopene, again others by stereospecific total synthesis, while some were isolated from plants.

Without a single exception each *cis* isomer exhibits maxima at wavelengths shorter than those of all-*trans*-lycopene.

Although the spectral behavior of diphenylpolyenes is very similar (*292, 522, 84*), this "rule" has no general validity. It does not apply either to

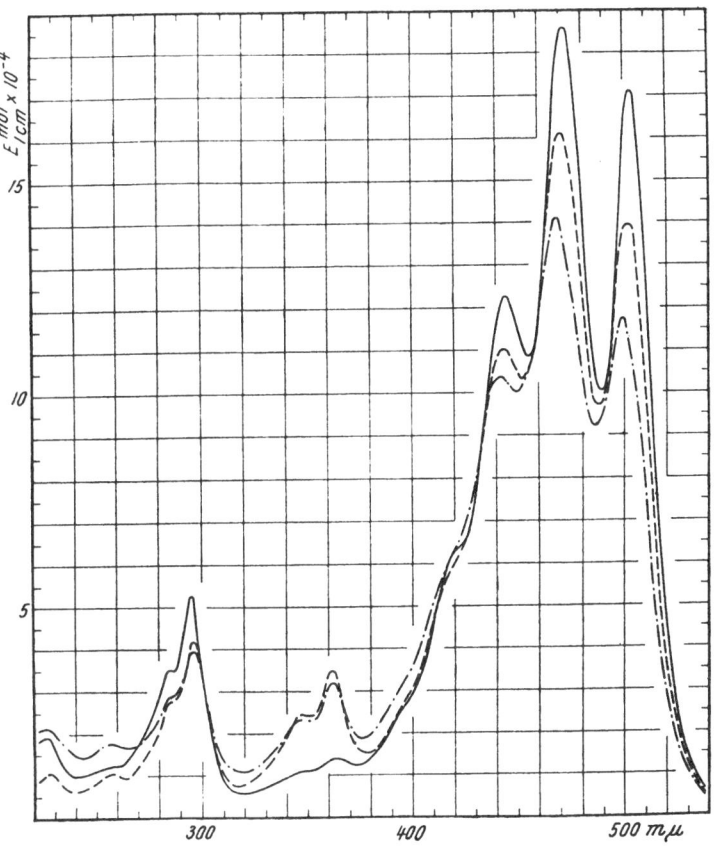

Fig. 6. Molecular extinction curves of lycopene, in hexane: ————, fresh solution of the all-*trans* compound; — — —, mixture of stereoisomers after refluxing in darkness for 45 min.; and — · — · —, after catalysis by iodine (*515*). [From: J. Amer. Chem. Soc. 65, 1940 (1943).]

vitamins A and retinenes (p. 126) or to dienes and some simple unbranched polyenes [NAYLER and WHITING (*333*), CROMBIE, HARPER and SMITH (*74*)]. Furthermore, according to HOLME, JONES and WHITING (*164*), the approximately coplanar all-*cis*-deca-2,4,6,8-tetraene shows *longer* wavelength maxima than the all-*trans* form. In our opinion (*517*) these observations should not disturb the general interpretation of carotenoid spectra which is based on a very great number of experimental facts. In the future, we hope, a more exact theory will cover all available spectroscopic observations.

While according to our concepts the spectral phenomena described are caused by spatial rearrangements about double bonds, an essentially different interpretation was proposed by SIMPSON (*406*). This author seems to assume that the formation of stereoisomeric carotenoids generally involves *cis* and *trans* configurations around single bonds of the aliphatic conjugated system ["*s-cis*" and "*s-trans*"; cf. (*330*)]. From the viewpoint of the organic chemist various experimental data would be in disagreement with such a theory.

(a) Iodine, the most successful agent in pertinent experiments, is a well-known catalyst that affects rearrangements especially around aliphatic double bonds. Most preferred *cis* forms obtained by means of iodine have been identified with those prepared by thermal rearrangements. (b) If some single bonds in carotenoids were *cis*, isomers might exist with very weak absorption in the fundamental region; such isomers have never been observed. (c) The presence of *cis* double bonds follows from infrared readings (p. 79). (d) Recently, numerous *cis* carotenoids have been prepared by partial reduction of triple bonds (p. 58). Although no *s*-isomerism could be involved here, the synthetic *cis* carotenoids show striking similarity with those obtained by rearrangement of all-*trans* forms; in certain instances identity has been established. (e) In the case of some simple polyenes, such as diphenylbutadiene, the number of the observed *cis* forms is identical with that calculated on the basis of *cis-trans* isomerism around double bonds. (f) According to the best available evidence, the contribution of *s-cis* conformations in acrolein and some methyl-substituted α,β-unsaturated aldehydes is of a low order of magnitude (1–2%) (*446, 120*).

The above comments refer to generalizations and should, of course, not imply that *s-cis* conformations are excluded from all aliphatic polyene systems. We refer, for example, to observations reported by WEEDON et al. (*3, 478*) concerning acyclic conjugated polyene diketones with bulky end groups. Some of their data could well be explained by the presence of *s-cis* bonds. Connected problems have been reviewed by WAIGHT and ERSKINE (*447*); see also p. 17.

3. Spectral Effect of *trans → cis* Isomerization in the Near Ultraviolet Region: the *cis*-Peak.

It was observed in 1943 in collaboration with POLGÁR (*524*) that, when an all-*trans* carotenoid undergoes stereoisomerization, causing the described changes in the visible region, simultaneously a new maximum, the so-called "*cis*-peak" grows out in the near-ultraviolet region, somewhere between 320 and 380 mμ (Figs. 3–6, pp. 26–29).

The "*cis*-peak effect" can be demonstrated, for example, by adding catalytic amounts of iodine and illuminating the solution or keeping it in diffuse daylight. As *Fig. 7* shows, no *cis*-peak appeared when an iodine-catalyzed lycopene solution was kept in darkness for an hour, but a light impulse as short as 5 seconds did produce a sizeable effect (*367*).

The appearance of the peak had been observed by earlier authors, after long standing of carotene solutions, but the reversible nature of this spectral change was overlooked and the phenomenon was ascribed to oxidation—a reasonable assumption at a time when nothing was known about stereoisomeric polyenes (*310, 318, 319*). In some other instances *cis*-peaks were represented as a feature of all-*trans* curves; evidently, non-intended partial stereoisomerization was responsible for such effects [cf. e. g. (*240, 106*)].

The best media for the observation of *cis*-peaks are non-polar solvents such as hexane, cyclohexane, or benzene. In carbon disulfide, which

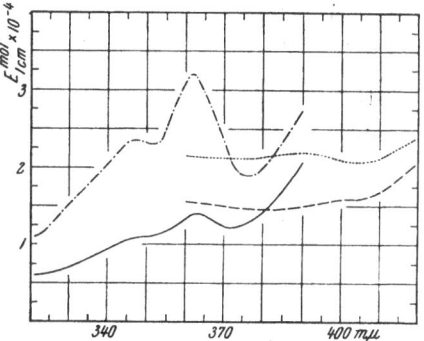

Fig. 8. Molecular extinction curves in the *cis*-peak region, in hexane and carbon disulfide solutions: ————, all-*trans*-lycopene, in hexane; and — · — · —, mixture of stereoisomers after catalysis by iodine; — — —, all-*trans*-lycopene, in CS₂; and · · · · · · · ·, after catalysis by iodine (*526*). [From: J. Amer. Chem. Soc. **67**, 108 (1945).]

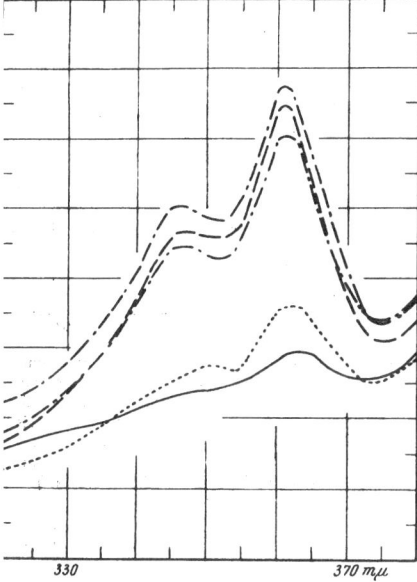

Fig. 7. Influence of illumination on the development of the *cis*-peak in an iodine-catalyzed solution of lycopene, in hexane: · · · · · ·, after 0 sec.; — — —, after 5 sec.; — — · — —, after 30 sec.; — · — · —, after 15 min. illumination; and ————, before the addition of iodine, without illumination (*367*). [From: J. Amer. Chem. Soc. **66**, 186 (1944)]

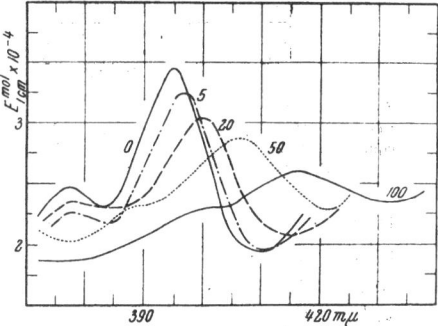

Fig. 9. Alteration of the molecular extinction coefficient in the *cis*-peak region of an iodine-catalyzed rhodoviolascin (spirilloxanthin) solution, in benzene, upon the addition of carbon disulfide (the figures on the curves indicate % CS₂ in the solvent) (*364*). [From: Arch. Biochemistry **5**, 243 (1944).]

decreases or destroys the fine structure also in the fundamental band, the curves are flat, although elevated, in the *cis*-peak region *(Figs. 8—9)* (*526*).

An extensive study of *cis*-peaks exhibited by stereoisomeric mixtures and by chromatographically pure, individual *cis* isomers has revealed the following regularities in carotenoid sets:

(a) The curves of all-*trans* carotenoids are flat in the *cis*-peak region but most *cis* isomers obtainable by direct rearrangement do show *cis*-peaks (*Table 5*).

(b) The *cis*-peaks of all observable members of a given stereoisomeric set are located at the same wavelength. Usually, some of the individual *cis*-peaks are higher and others are lower than

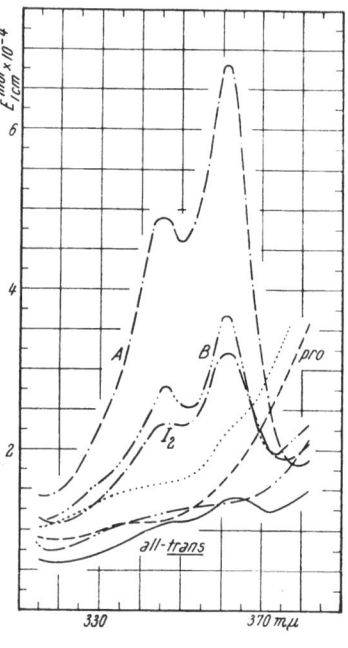

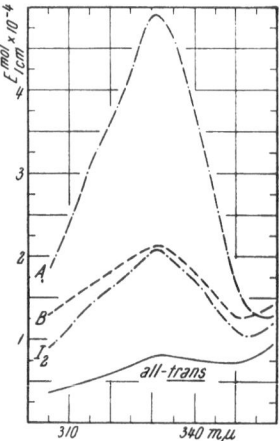

Fig. 10. Molecular extinction curves of some lutein stereoisomers in the *cis*-peak region, in hexane. I_2 indicates the equilibrium mixture obtained upon catalysis by iodine (*525*). [From: J. Amer. Chem. Soc. 66, 137 (1944).]

Fig. 11. Molecular extinction curves of some lycopene stereoisomers in the *cis*-peak region, in hexane. I_2 indicates the equilibrium mixture obtained upon catalysis by iodine, and — ·· — ·· —, an unnamed crystallizable stereoisomer (*525*). [From: J. Amer. Chem. Soc. 66, 137 (1944).]

that of the stereoisomeric equilibrium mixture; on catalysis by iodine the former peaks decrease and the latter increase to reach the equilibrium value (*Figs. 10–11*).

(c) In each stereoisomeric set only a few of the observed stereoisomers have high *cis*-peaks and these are mainly responsible for the peak observed in the stereoisomeric equilibrium mixture; in some instances a single *cis* form must be credited for the major part of the effect (*Table 6*, p. 34).

(d) In the curves of C_{40}-carotenoids that possess about 10—11 conjugated double bonds, the *cis*-peak has a well-defined location: the wavelength difference between its position and the longest wavelength maximum of the all-*trans* compound is practically a constant, viz. 142 mμ

Table 5. Examples of the *cis*-Peak Intensities in Some Stereoisomeric Sets (within the respective sets the stereoisomers are listed in the sequence of decreasing adsorption affinities).

Stereoisomeric set	Member of the set	Difference between the longest wavelength maximum of the *cis* isomer and that of the all-*trans* form (mμ)	Molecular extinction coefficient at *cis*-peak $E^{mol.}_{1\ cm.} \times 10^{-4}$	Difference between $E^{mol.}_{1\ cm.} \times 10^{-4}$ for member of the set and for all-*trans* form at *cis*-peak	References
α-Carotene (in hexane)	Neo U	5.5	1.2	0.4	(524)
	Neo V	11.5	1.1	0.3	
	Neo W....	6.5	1.6	0.8	
	Neo X....	13.5	2.7	1.9	
	All-*trans*...	0	0.8	0	
	Neo A	8.5	3.8	3.0	
	Neo B	10.5	3.8	3.0	
	Neo C	4.5	4.5	3.7	
β-Carotene (in hexane)	Neo U	5	1.3	0.5	(524)
	Neo V	13.5	0.8	0	
	All-*trans*...	0	0.8	0	
	Neo B	10.5	3.4	2.6	
	Neo E	8.5	3.4	2.6	
γ-Carotene (in hexane)	All-*trans*...	0	0.95	0	(515, 526)
	Neo U	5	1.4	0.4	
	Pro-γ	31	1.3	0.35	
Lycopene (in hexane)	All-*trans*...	0	1.4	0	(515, 524)
	Neo A	5	6.8	5.4	
	Neo B	8	3.7	2.3	
	Unnamed crystalline isomer ...	28	1.3	— 0.1	
	Prolycopene	34	1.6	0.2	
	Another poly*cis* form	38.5	2.2	0.8	
Cryptoxanthin (in hexane)	Neo U	5	1.5	0.3	(512)
	All-*trans*...	0	1.2	0	
	Neo A	6.5	4.2	3.0	
	Neo B	4.5	4.5	3.3	
Cryptoxanthin (in benzene)	Neo U	5	1.7	0.7	(512)
	All-*trans*...	0	1.0	0.7	
	Neo A	6.5	3.4	2.4	
	Neo B	4.5	4.6	3.6	
Zeaxanthin (in benzene)	Neo A	5.5	4.4	3.7	(512)
	Neo B	5.5	2.4	1.7	
	Neo C	8.5	3.9	3.2	
	All-*trans*...	0	0.7	0	
Lutein (in benzene)	Neo A	6	4.9	4.1	(524)
	Neo B	7	2.1	1.3	
	All-*trans*...	0	0.8	0	
Capsanthin (in benzene)	Neo A	6	4.4	3.4	(367)
	Neo B	6	2.7	1.6	
	Neo C	10.5	1.9	0.8	
	All-*trans*...	0	1.1	0	

Zechmeister, Carotenoids.

($\pm$ 2 mμ) in hexane solution and a little larger (about 145–146 mμ) in benzene. This statement is valid for a large number of carotenoids whose molecules contain, respectively, aliphatic, hydroaromatic, or aromatic terminal groups. In the case of shorter conjugated systems the distance mentioned is markedly reduced *(Table 7*; for the vitamin A and retinene sets see p. 126).

Table 6. Individual Stereoisomers Responsible for the Major Part of the *cis*-Peak Effect Observed in Equilibrium Mixtures upon Iodine Catalysis of All-*trans* Compounds.

Stereoisomeric set	Name of *cis* form	Approximate percentage of the *cis* form in the equilibrium	Approximate percentage of the total *cis*-peak effect caused by the *cis* form	References
α-Carotene	Neo B	13	55	*(525)*
β-Carotene	Neo B	25	75	*(365)*
Lycopene	Neo A	30–40	95	*(515)*
Lutein..............	Neo A	17	70	*(536)*
Cryptoxanthin.......	Neo A	23	60	*(512)*
Zeaxanthin..........	Neo A + B	30	90	*(512)*
Capsanthin..........	Neo A	20	80	*(367)*

(e) The spectral curves of the naturally occurring poly*cis* compounds, for example prolycopene (p. 68), are as flat in the *cis*-peak region as those of all-*trans* carotenoids; a peak appears, however, on catalysis by iodine.

(f) The presence of a hindered *cis* double bond does not exclude the presence of a *cis*-peak.

Example. Hindered *cis* lycopenes (p. 99).

(g) No distinct *cis*-peak was observed in the case of *cis retro* carotenoids which possess a long conjugated system (e. g. *retro*-dehydrocarotene, *Fig. 12*, p. 36), possibly because the peak was "covered up" by the broad main band *(537)*. Certain *retro* compounds with shorter chromophores may show the peak *(191)*.

(h) Some of the observed *cis*-peaks have fine structure but others do not. Although such differences cannot be interpreted at the present time, the following empirical statements can be made. If the molecule of a C$_{40}$-carotenoid (hydrocarbon or alcohol) contains two hydroaromatic end groups, the *cis*-peak is void of fine structure (α-carotene, β-carotene, 3,4-dehydro-β-carotene, 3,4,3',4'-bisdehydro-β-carotene, 16,16'-*homo*-β-carotene, cryptoxanthin, zeaxanthin, canthaxanthin, etc.). Nor does fine structure appear in the semi-aliphatic set of γ-carotene or in capsanthin but it is observed in several entirely aliphatic sets (lycopene, rhodoviolascin, phytofluene, etc.). β-Carotene diepoxide does show fine structure in the *cis*-peak region, in contrast to the monoepoxide. When the terminal groups are unmethylated aromatic rings, fine structure in the *cis*-peak region appears (1,18-diphenyl- or -dinaphthyl-3,7,12,16-tetra-methyloctadecanonaene, aromatic polyene azines, etc.); this is not the case, however, in the presence of several aromatically bound CH_3-groups (renieratene, isorenieratene, p. 105).

Table 7. Examples of the Position of the *cis*-Peak in Various Stereoisomeric Sets.

Name	Number of conjugated C=C double bonds	Position of *cis*-peak (mμ)	Distance between *cis*-peak and		References
			longest wave length max.	λ_{max}	
			of the all-*trans* form (mμ)		

In hexane or petroleum ether solution:

Name	Number of conjugated C=C double bonds	Position of *cis*-peak (mμ)	longest wave length max.	λ_{max}	References
Rhodoviolascin (spirilloxanthin)	13	384	143	109	(364)
3,4,3',4'-Bisdehydro-β-carotene	13	368	—	103	(191)
Lycopene	11	362	141	111	(524, 197)
Canthaxanthin	11 + 2 C=O	356	—	110	(123)
Capsanthin	10 + 1 C=O	355	143	109	(367)
16,16'-*Homo*-β-carotene*	12	355	141	111	(183)
3,4-Dehydro-β-carotene	12	352	—	109	(204)
4-Hydroxy-γ-carotene	11	350	141	110	(49)
γ-Carotene	11	349	143	112	(524)
β-Carotene	11	338	142	114	(524, 185)
Cryptoxanthin	11	339	141	113	(512, 202)
Physaliene	11	338	141	113	(524)
Isozeaxanthin	11	337	141	114	(205)
Zeaxanthin	11	336	144	116	(201, 207)
Neurosporene	9	332	136	118	(511)
α-Carotene	10	331	143	114.5	(524, 192)
Lutein	10	331	143	114.5	(524)
Antheraxanthin**	10	331	141	115	(432)
5,6-Dihydro-β-carotene	10	331	143	114	(524)
4-Hydroxy-α-carotene	10	330	144	115	(49)
β-Carotene monoepoxide	10	330	145	116	(440)
5,6-Dihydro-α-carotene	9	329	140	109	(524)
β-Carotene diepoxide	9	328	142	111	(440)
Dimethylcrocetin	7 + 2 C=O	314	134	108	(188)
ζ-Carotene	7	296	129	104	(511)
Diphenyloctatetraene	4 + 2 C_6H_5	283	111	89	(522)
Diphenylhexatriene	3 + 2 C_6H_5	268	101	83	(292)
retro-Dihydro-C_{19}-aldehyde	5 + 1 C=O	267	135	114	(191)
Phytofluene	5	260	107	88	(350)
2,7-Dimethyl-octa-2,4,6-triene-1,8-dial	3 + 2 C=O	234	102	85	(182)

* In ether.

** In alcohol.

(continued on next page)

(Table 7, continued.)

Name	Number of conjugated C=C double bonds	Position of *cis*-peak (mμ)	Distance between *cis*-peak and		References
			longest wave length max. of the all-*trans* form (mμ)	λ_{max}	
In benzene solution:					
Rhodoviolascin (spirillo-xanthin).............	13	395	151	115	*(364)*
1,18-Diphenyl-3,7,12,16-tetramethyl-octa-decanonaene	9 + 2 C$_6$H$_5$	378	142	109	*(106)*
Canthaxanthin	11 + 2 C=O	366	—	114	*(123)*
Methylbixin	9 + 2 C=O	363	145	112	*(509, 182)*
Capsanthin	10 + 1 C=O	362	(146)	122	*(367)*
Renieratene............	9 + 2 C$_6$H$_5$	362	145	114	*(487–489)*
Cryptoxanthin	11	348	145	115	*(512, 59)*
Zeaxanthin	11	348	144	114	*(512)*

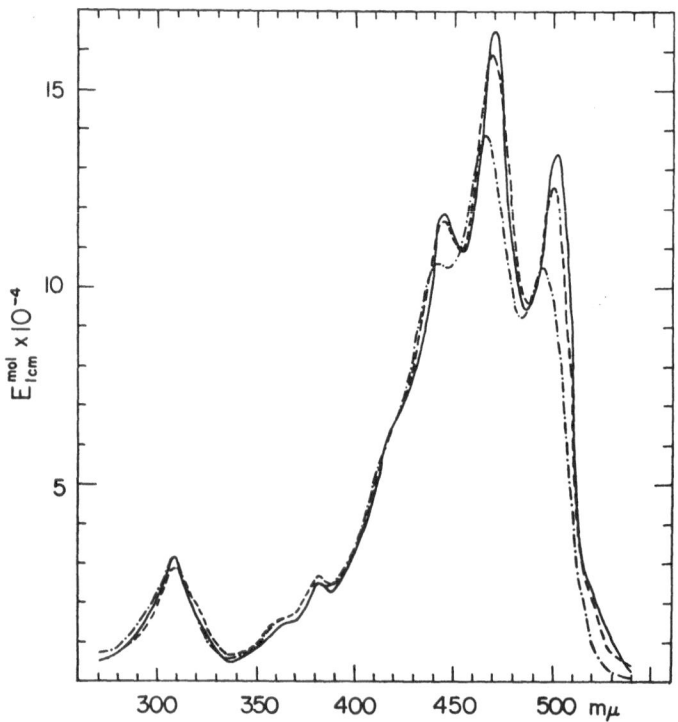

Fig. 12. Molecular extinction curves of *retro*-dehydrocarotene, in hexane: ————, fresh solution of the all-*trans* compound; — — —, mixture of stereoisomers after refluxing for 1 hour, in darkness; and — · — · — after catalysis by iodine, in light *(537)*. [From: J. Amer. Chem. Soc. **75**, 5341 (1953).]

(i) The presence or absence of fine structure in the fundamental band does not determine the corresponding situation in the *cis*-peak region.

(j) Whenever the *cis*-peak region exhibits fine structure, the curve has a characteristic shape; it shows a lower maximum (or shoulder) and a higher maximum, the latter being located nearer longer wavelengths.

(k) LOEB, BROWN and WALD (*288*) have shown for central-mono*cis*-β-carotene that the spectral effect of cooling to the temperature of liquid nitrogen parallels the shift observed in the main band: the *cis*-peak is shifted by 14 mμ toward longer wavelengths and elevated by 9% (cf. p. 44).

4. A Simple Theoretical Interpretation of Spectral Phenomena, Especially of the *cis*-Peak Effect.

This Section will be restricted to a few features that are of immediate interest to the organic chemist and illustrate the close relationship between light absorption and the over-all shape of the polyene molecule.

For recent surveys cf. CROMBIE (*73*), WYMAN (*486*), H. KUHN (*254–256*), BRAUDE and WAIGHT (*41*). (The latter authors term the fundamental and the *cis*-peak bands, respectively, "full-chromophore" and "half-chromophore" bands.)

Molecular orbital methods will not be considered here. They were applied to polyenes by COULSON (*72*); several possible theoretical approaches were outlined by MULLIKEN (*331*). The free-electron molecular-orbital method was discussed recently by PLATT (*363*) and especially by H. KUHN (*254, 255*).

The writer would recommend caution in the theoretical interpretation of minor bands because of possible spontaneous stereoisomerization in solutions and the resulting appearance of minor *cis* bands which are not immediately recognizable as such.

In 1939 there appeared several theoretical treatments that have profoundly influenced further research. We mention PAULING's contributions (*347, 348*) and the extensive paper by LEWIS and CALVIN (*284*) on the color of organic compounds in which the bands of "partial oscillation" are also treated. An important theoretical study of electronic transitions in molecular spectra of conjugated polyenes including carotenoids was presented by MULLIKEN (*329*).

According to MULLIKEN the shape of the conjugated polyene molecule strongly affects the distribution of absorption intensity. The more elongated the molecule, the greater the total intensity and the greater the preponderance of intensity in the longest wavelength transition; such $N \rightarrow V_1$ transitions are polarized approximately along the long axis of the molecule. "Molecules which for any reason may be more *cis*-like, should have weaker $N \rightarrow V_1$ spectra." This postulate is illustrated by the substitution of a terminal furyl group for methyl, whereby the intensity of the fundamental band increased much less than upon the introduction of two aliphatic conjugated double bonds. The conjugated system is bent into the furan ring, in a "*cis*-like" manner.

In part based on these theoretical treatments, the following quasi-classical interpretation of our spectroscopic observations was presented in 1943 by PAULING, LeROSEN, SCHROEDER, POLGÁR, and the writer (*515*).

The three regions in carotenoid spectra which are of practical importance are: (a) the region of extraordinarily strong extinction in the visible region (fundamental or K band or λ_1-band), (b) the *cis*-peak region in the near-ultraviolet (first overtone band or λ_2-band), and (c) a region in the farther ultraviolet (second overtone or λ_3-band). These bands correspond to transitions from the normal electronic state to excited states. The fundamental band results from the transition $0 \rightarrow 1$, the *cis*-peak from $0 \rightarrow 2$, and the second overtone from $0 \rightarrow 3$. (For example, the three corresponding maxima of lycopene in hexane solution are located at 471, 361, and 296 mμ; Fig. 6, p. 29.) These electronic levels may be discussed in terms of the conventional structure $\dots\diagup\diagdown\diagup\diagdown\diagup\diagdown\diagup\dots$, and a great number of ionic structures, such as

The conventional structure of alternating double and single bonds makes a most important contribution to the normal state, whereas the ionic structures contribute in the main to the excited states.

The three transitions mentioned correspond to oscillation of the electric charge along the unsaturated chain. Following LEWIS and CALVIN (*284*) we may compare these with the classical modes of vibration of mobile "unsaturation" electrons of the conjugated system along the chain. The observed bands can be correlated with the following modes of oscillation (*515*).

(a) Fundamental band: The electrons tend to concentrate first near one end and then near the other end of the chain; this simple oscillation would, according to classical electromagnetic theory, result from the absorption of light of the proper frequency because of interaction of the electric vector of the light and the regularly reversing electric dipole moment of the molecule. This mode of oscillation of the charge corresponds to the transition $0 \to 1$.

(b) Band in the *cis*-peak region: This results from the oscillation of the electrons from the two ends of the conjugated system toward the middle and from the middle toward the two ends (transition $0 \to 2$).

(c) Next band in the farther ultraviolet: This involves concentration of the electrons alternately in the first and third and the second and fourth quarters of the conjugated system ($0 \to 3$).

The intensity of an absorption band is proportional to the square of the corresponding dipole moment, and hence essentially to the square of the length of the system (*347, 329, 330*). The maximum intensity of the fundamental band is shown by the all-*trans* molecule (Model I, Fig. 13, next page).

All *cis* isomers which have a vertical plane of symmetry, such as Models II and V (Fig. 13) have a distance between the ends of the conjugated system smaller than the all-*trans* isomer by the factor cos α, with $\alpha = 27° 22'$ (carbon bond angle along the chain $= 125° 16'$). The fundamental band for these isomers should then be approximately 80% as intense for the all-*trans* form since $\cos^2 \alpha = 0.80$.

According to BRAUDE and WAIGHT (*41*) one should calculate rather with the value $120°$; hence, $\cos^2 30° = 0.75$.

The intensities for all mono*cis* isomers should lie between the extremes of the all-*trans* and central-mono*cis* forms. (In comparing intensities, essentially the ratio of the extinguished areas rather than the height of peaks should be considered.)

The fact that this agrees with observation provides support for the assumption that the single bonds of the conjugated system have essentially the *trans* configuration. If some single bonds also were *cis*, stereoisomers might exist with very small absorption in the fundamental region (cf. p. 30).

The nature of the $0 \to 2$ oscillation, which was discussed by MULLIKEN (*329*) is such that it gives rise to no dipole moments and to no absorption band for the all-*trans* molecule or any other molecule with a center of symmetry. In contrast, a central-mono*cis* form (such as Model II, Fig. 13) does have a dipole moment for the transition $0 \to 2$, because of its bent shape (*515*). This dipole moment is perpendicular to the straight line, connecting the ends of the conjugated system, instead of parallel to it. Model II would show the strongest *cis*-peak of all. Certain other isomers such as Model III (Fig. 13) would also have *cis*-peaks but of smaller intensity. As a rough approximation the intensity

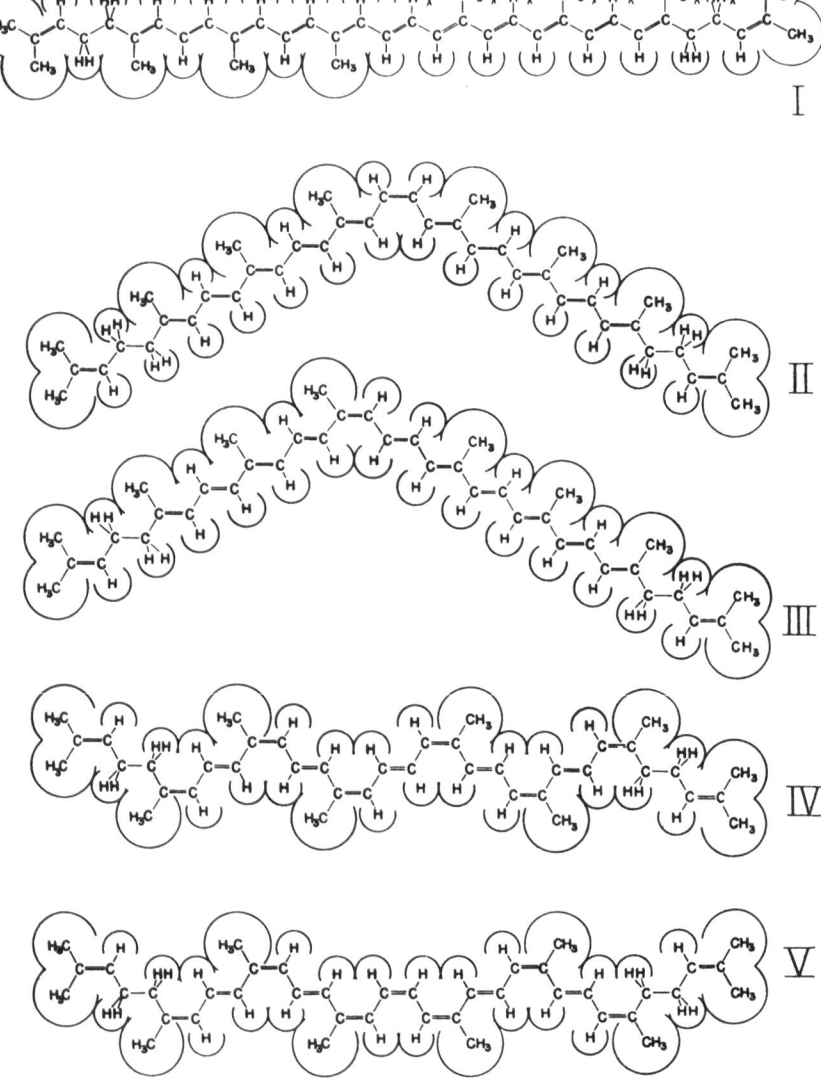

Fig. 13. Models of some lycopene stereoisomers: I, all-*trans* form; II, central-mono*cis*-lycopene; III, next-to-central-mono*cis*-lycopene; IV, a hexa*cis*-lycopene; and V, a hepta*cis*-lycopene in which all sterically unhindered double bonds are in the *cis* configuration. The bond angles, relative bond distances and VAN DER WAALS radii correspond approximately to the accepted values (515). [From: J. Amer. Chem. Soc. 65, 1940 (1943).]

of the *cis*-peak can be taken proportional to the square of the distance between the center of the conjugated system and the mid-point of the straight line between its two ends. Accordingly, only a few steric forms can have *cis*-peaks approaching the intensity of Model II.

Recently, these concepts have been confirmed experimentally by ECKERT and H. KUHN (*100*) who investigated the dichroism obtained by stretching a polyethylene film that contained central-mono*cis*-β-carotene in solid solution. The pigment molecules were oriented preferentially parallel to the direction of stretching. It was found that the sign of the dichroism was positive in the main band but negative at *cis*-peak. The linearly polarized light was more strongly absorbed when the electric vector was parallel to the direction of stretching than when it was perpendicular to it (see also *392*).

It should be emphasized that the above discussion is necessarily of qualitative nature in the sense that in a given stereoisomeric set it has

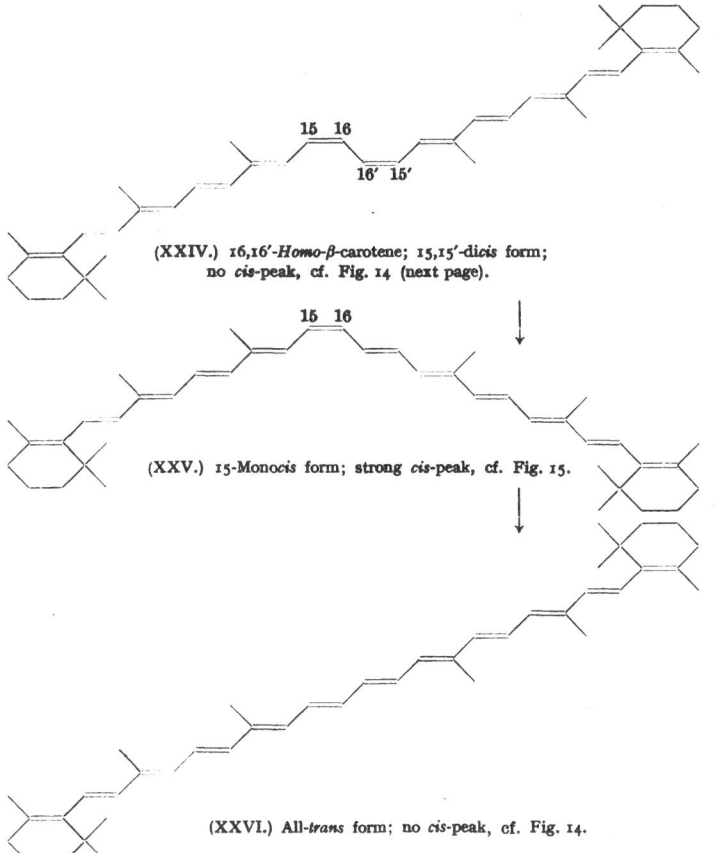

(XXIV.) 16,16'-*Homo*-β-carotene; 15,15'-di*cis* form; no *cis*-peak, cf. Fig. 14 (next page).

(XXV.) 15-Mono*cis* form; strong *cis*-peak, cf. Fig. 15.

(XXVI.) All-*trans* form; no *cis*-peak, cf. Fig. 14.

not yet been possible to predict on theoretical grounds the absolute value of the maximum possible (or any other) *cis*-peak intensity. Luckily, as will be shown later, this gap has been narrowed by measuring the *cis*-peaks of totally-synthetic central-mono*cis*-carotenoids (pp. 77, 81).

Let us now, in a mental experiment, start from an all-*trans* carotenoid and bend the molecule in the center by forming there a *cis* double bond, and then either slide the *cis* configuration along the chain or introduce more *cis* double bonds. The initially very high *cis*-peak will decrease in both instances and disappear as the molecule straightens out. These

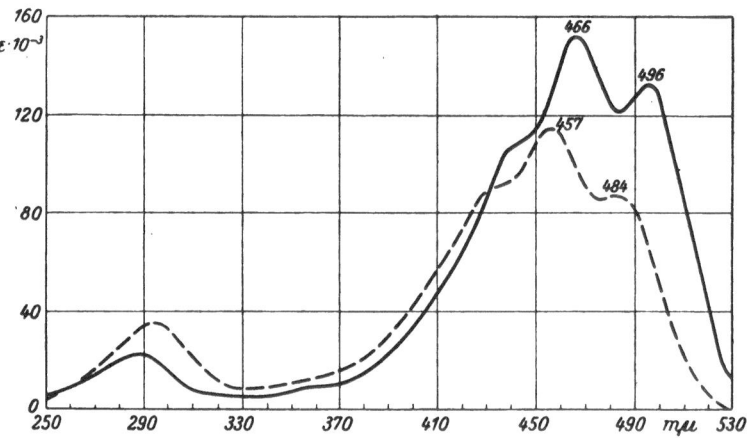

Fig. 14. Molecular extinction curves of 16,16′-*homo*-β-carotene, in ether: — — —, middle-di*cis* form; and ————, all-*trans* form; according to Inhoffen et al. (*183*). [From: Liebigs Ann. Chem. 573, 1 (1951).]

postulates are in good agreement with numerous experimental data, especially with the absence of *cis*-peaks from the curves of poly*cis* carotenoids.

A clear experimental demonstration of the modification of the *cis*-peak during the process, di*cis* (XXIV) → middle-mono*cis* (XXV) → all-*trans* (XXVI, p. 41), was given recently by Inhoffen (*183*). The absolute configuration of the di*cis* compound used was established by total synthesis and, as expected, did not cause any *cis*-peak. On two-step rearrangement of the 15,15′-di*cis* form of the 16,16′-β-carotene homolog $C_{48}H_{58}$ first an intense *cis*-peak appeared which disappeared during the second step *(Figs. 14–15)*. Such well-defined, stepwise isomerization experiments might well become important tools in future stereochemical research.

Recently, the empirical relationships of the minor bands, including *cis*-peaks in polyene spectra have been subjected to a detailed analysis by Dale (*79*) who has stated the following simple rule: In case of an aliphatic polyene containing n conjugated double bonds, the wavelength maximum of any minor band (λ_s) will lie very close to that of the main band of a corresponding polyene with n/s conjugated double bonds. This rule seems to apply also to poly*cis* carotenoids and numerous other polyenes.

The observed characteristics of certain *retro* carotenoid spectra (very sharp fine structure in the main band; great similarity between *trans* and *cis* extinction curves in the main band but lack of a conspicuous

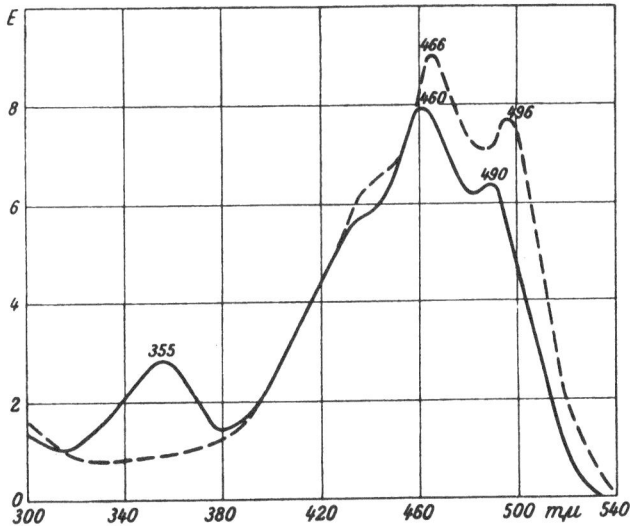

Fig. 15. Molecular extinction curves of middle-monocis-16,16'-*homo*-β-carotene, in ether: ————, fresh solution; and — — —, after exposure to diffuse daylight for 5 hours; according to INHOFFEN et al. (*183*).
[From: Liebigs Ann. Chem. 573, 1 (1951).]

cis-peak effect; cf. Fig. 12, p. 36) as compared to those of *normal* carotenoids is explained by a spatial conflict between the *gem*-dimethyl group and the opposing chain-hydrogen in *normal* carotenoids whose cyclohexene ring is somewhat twisted. In contrast, the ring is planar and the whole conjugated system very nearly coplanar in *retro* carotenoids (*79*).

DALE (*80*) has presented a theoretical basis of the above findings by analyzing the free-electron model of polyenes. Furthermore, he has shown that the minor bands are overtones also in polyenynes and, especially in dehydro-carotenoids, not due to isolated partial chromophores [connected papers (*81–83*)]. Some quantum-mechanical aspects of the polyene spectra, based on the free electron gas theory have recently been discussed by H. KUHN (*254, 255*). He has presented a theoretical explanation of DALE's selection rule, according to which only the electrons of the outermost π electron shell are excited by light in polyenes.

Considering the nature of the *cis*-peak effect it is hardly necessary to stress that this phenomenon is by no means restricted to carotenoid pigments. Since the intensity of the peak is determined by the bent molecular shape, the *cis*-peak must be considered as an important characteristic of all such aliphatic multiconjugated systems that can be subjected to geometrical isomerization. Accordingly, *cis*-peaks have been observed in the following classes of compounds (among others): Acidic carotenoids (*490*); lower-molecular weight isoprenic systems, e. g. methylbixin (*509*) (p. 112), vitamin A (*167*) (p. 127), retinene (*167*) (p. 127); non-isoprenic aliphatic systems with aromatic end groups: diphenylhexatriene (*292*) (p. 171), diphenyloctatetraene (*522*) (p. 176); renieratene (*487, 489*) (p. 105); polyphenyl-polybutadienes (*99*) (p. 168); polyene azines (*84*) (p. 195); and entirely aliphatic, non-isoprenic systems: corticrocin (*105*), unbranched *cis*-polyenynes (see e. g. *39*), etc. In contrast, when a *trans* → *cis* rearrangement does not alter the straight overall shape of the aliphatic system, no *cis*-peak appears; example, dibiphenylene-hexatriene (*299*) (p. 187).

As mentioned *cis*-peaks have also been observed in the solid solution state (p. 41) (*100, 275, 277*) as well as in the glassy state. β-Carotene "glasses" were obtained by ROSENBERG (*392*) by fusing the crystals and solidifying the melt. He found that the *trans* → *cis* rearrangement process was important in determining the photoconduction phenomena in the glass. The observed photoconductive excitation spectrum was very different from the optical extinction curve of all-*trans*-β-carotene solutions and showed maximum effect in the *cis*-peak region.

5. Spectra at Extremely Low Temperatures.

The first observation in this interesting field was made in 1935 by HAUSSER, KUHN and SEITZ (*153*) who found that the UV spectra of α,ω-diphenylpolyenes, some polyene-carboxylic acids and lycopene underwent conspicuous changes on cooling their solutions in liquid nitrogen (— 185° to — 195°): in the fundamental region the curves became much sharper, some new maxima appeared and the extinction values increased far beyond those observed at room temperature. While only all-*trans* compounds could be investigated at HAUSSER's time, quite recently WALD and his collaborators have published an important study into the spectral behavior of geometric isomers of some carotenoids, vitamin A and retinene (p. 129) [JURKOWITZ (*220*), LOEB, BROWN and WALD (*288, 459*)].

On cooling with liquid nitrogen, the all-*trans*-β-carotene spectrum shows the following changes: λ_{max} is shifted by 17.5 mμ towards longer wavelengths, the fine structure is accentuated with two new maxima appearing; the longest wavelength maximum becomes the highest (*Fig. 16*).

The intensity of the middle maximum is then 1.34 times higher than at 20°. The sterically unhindered central-mono*cis*-β-carotene (p. 81) behaves similarly; the *cis*-peak is displaced by 14 mμ towards longer wavelengths but its intensity is increased by only 9%.

Dramatic changes take place upon cooling the sterically hindered 11,11′-di*cis*-β-carotene (Fig. 16): instead of the degraded spectrum

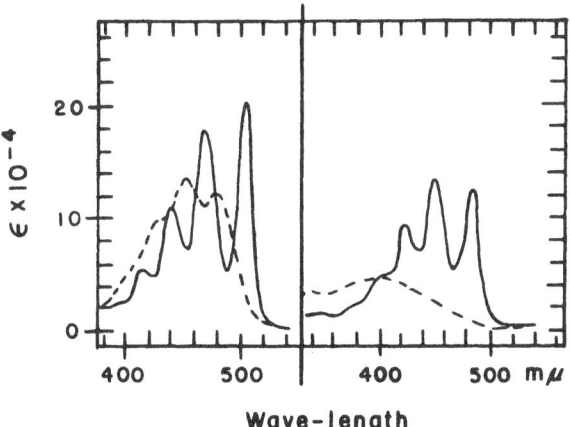

Wave-length

Fig. 16. Influence of extreme low temperature on the molecular extinction curves of all-*trans*-β-carotene (left) and of the hindered isomer 11,11′-di*cis*-β-carotene (right) in ether-isopentane-alcohol: — — —, at room temperature; and —————, at — 185° to — 195°. According to LOEB, BROWN and WALD (*288*). [From: Nature **184**, 614 (1959).]

extensive fine structure appears, the position of λ_{max} is shifted by 49.5 mμ towards longer waves and its intensity becomes 2.9 times higher than that observed at room temperature; the extinguished area is increased by 80%. Spectroscopically speaking, it seems as if the cooling would "relieve" the hindrance caused by the conflict between a H-atom and CH_3-group at a hindered *cis* double bond (or bonds).

All-*trans*-lycopene shows analogous spectral changes.

A theoretical discussion of these phenomena has been presented by WALD (*288, 459*).

For *infrared* spectra see pp. 79, 91; and for *nuclear magnetic resonance* spectra, WEEDON et al. (*22, 24, 21*).

Photoconduction activation energies of all-*trans* and central-mono*cis*-β-carotenes: ROSENBERG (*392*).

V. Preparation of *cis* Carotenoids by Direct Rearrangement of the All-*trans* Form.

Cis carotenoids have been obtained by stereoisomerization of all-*trans* pigments, by total synthesis, and by isolation from plants or animals.

In general, the rearrangement methods afford quasi-equilibria or steady-state mixtures that contain *cis* isomers and a substantial portion of unchanged all-*trans* pigment. The composition of the mixtures is revealed on chromatographic resolution followed by spectral characterization and estimation of the individual *cis* isomers. The result will depend on the nature of the treatment; thus, thermal and photochemical methods may afford different equilibria.

Examples. (a) Very little neolutein B was obtained upon refluxing the all-*trans* solution but it appeared in substantial quantities upon catalysis by iodine, in light (*536*). (b) One of the *cis* β-carotene diepoxides was not contained in equilibrium mixtures prepared by refluxing or illumination in the absence of catalysts, but it was formed to the extent of 25% in the presence of iodine (*440*).

1. Thermal Methods of *cis-trans* Isomerization.

a. Spontaneous Rearrangements in Solution at Room Temperature. This usually slow process starts on dissolution of carotenoid crystals or elution of adsorbates and proceeds at rates which depend on structure, configuration, solvent, temperature, and light.

Table 8. Examples of the Spontaneous Rearrangement of *cis* Carotenoids in Benzene Solution.

(The neo-β-carotene B solutions were sealed in dark brown ampoules; the others were exposed to weak diffuse daylight.)

Compound	Temperature	Duration	Fraction stereoisomerized (%)	References
Neo-β-carotene B	— 2°	60 days	5–6	(*55*)
Neo-β-carotene B	20°	1 day	5	(*55*)
Neo-β-carotene B	20°	14 days	20	(*55*)
Neo-β-carotene B	20°	60 days	47	(*55*)
Neocapsanthin A or B ...	20°	1 day	50	(*503*)
Neorhodoviolascin A	20°	9 min.	55	(*364*)

The photolabile central-mono*cis*-β-carotene remained unchanged for a long time at — 5°, in darkness (*187*).

Examples. (a) According to CARTER and GILLAM (*55*) the extent of stereo-isomerization of all-*trans*-β-carotene in benzene-petroleum ether amounted to 11% in 49 days at 20° but only to 3–4% in 90 days at — 2° (in darkness, protected from air). (b) At room temperature, the following fractions of all-*trans* carotenoids underwent stereoisomerization in benzene or light petroleum solution within a day, in part in diffuse daylight (*492, 535, 526, 503, 364*): α-, β-, γ-carotene, crypto-xanthin, and capsanthin, 1–2%; zeaxanthin, 4–5%; capsorubin, 8%; lycopene, 10%; rhodoviolascin (spirilloxanthin), 23%.

When working with carotenes the spontaneous isomerization can be neglected in short experiments, but the lability of some pigments such as all-*trans*-rhodoviolascin is a disturbing factor; chromatography revealed that even fresh solutions of crystals contained appreciable amounts of *cis* isomers (*364*).

The few *cis* carotenoids whose spontaneous *cis* → *trans* rearrangement has been studied in some detail show substantial individual differences, also depending on the conditions *(Table 8)*.

b. cis-trans Isomerization in Refluxed Solutions. When a dilute petroleum ether, hexane, cyclohexane, or benzene solution of an all-*trans* carotenoid is refluxed (preferably in darkness or semi-darkness) the stereoisomeric equilibrium is reached within 10–60 minutes. *Table 9* gives some information on the composition of the resulting solutions; and in *Table 10* the behavior of all-*trans*-β-carotene is compared with that of its two main isomers (pp. 48, 50).

Under the conditions of refluxing many neo forms are labile. SAVINOV (*397*), however, was able to prepare neo-β-carotene B in good yield by keeping all-*trans* solutions at 100° in sealed tubes.

Some poly*cis* carotenoids are as thermostable as the corresponding all-*trans* forms (p. 68).

c. cis-trans Isomerization by Melting Crystals. The crystals are sealed in vacuo or under an inert gas in a narrow tube which is kept in a bath a few degrees above the melting point for 1–15 min. The tube is then rapidly cooled in ice-water, the substance is dissolved in cold solvent and chromatographed immediately (*365, 525, 367, 282*). No true equilibria are expected under these conditions but more or less stable steady-state mixtures that usually contain also some irreversibly formed product. The partial cleavage process may be checked to a certain extent when the melting point is depressed by the addition of naphtha-lene (*512*).

The method is useful for the isolation of some otherwise unavailable steric forms, the more so since *cis* isomers may predominate over the *trans* form in the melt. In exceptional instances small amounts of sterically hindered *cis* carotenoids have also been obtained (pp. 91, 194).

Table 9. Examples of all-*trans* → *cis* Isomerization of Carotenoids in Refluxed Solutions.

(In the column "Ratio of stereoisomers" the first figure designates the top zone in the chromatogram; the values referring to unchanged all-*trans* forms are italicized; trace amounts are neglected.)

Type of compound	Name	Solvent	Duration (min.)	Observed number of *cis* forms	Extent of isomerization in the recovered pigment (%)	Ratio of stereoisomers in the recovered pigment	References
Hydrocarbons	α-Carotene	petr. ether b. p. 60—70°	30	2	8	4 : *92* : 4	(525)
	β-Carotene	petr. ether b. p. 60—70°	60	> 4	14	4 : *86* : 8 : 1 : 1	(365)
	γ-Carotene	petr. ether b. p. 60—70°	30	≫ 4	27	*73* : 6 : 5 : 8 : 8	(526)
	5,6-Dihydro-α-carotene	petr. ether b. p. 60—70°	30	> 4	36	3 : 27 : *64* : 6	(366)
	5,6-Dihydro-β-carotene	petr. ether b. p. 60—70°	30	3	11	2 : *89* : 9	(366)
	3,4-Dehydro-α-carotene	hexane	45	3	10	2 : *90* : 5 : 3	(221)
	retro-Dehydrocarotene	hexane	60	> 4	36	*64* : 28 : 5 : 2 : 1	(537)
	Neurosporene	hexane	60	8	16	*84* : 16	(300)
Alcohols	Cryptoxanthin (3-hydroxy-β-carotene)	ligroin b. p. 120°	30	2	38	*62* : 32 : 6	(504, 572)

	4-Hydroxy-α-carotene	hexane	30	3	5	95 : 2 : 2 : 1	(48)
	Zeaxanthin (3,3'-dihydroxy-β-carotene)	benzene	30	> 2	30	24 : 6 : 70	(536, 512)
	Lutein (3,3'-dihydroxy-α-carotene)	benzene	30	> 2	13	13 : 87	(536)
	5,6-Dihydroxy-5,6-dihydro-lycopene	benzene	30	> 4	4	96 : 1 : 2 : 1	(48)
Esters	Physaliene (zeaxanthin dipalmitate)	petr. ether b. p. 70–80°	60	1	42	58 : 42	(504)
	Capsanthin dipalmitate	petr. ether b. p. 70°	30	2	36	64 : 36	(503)
Dimethoxy-hydrocarbons	Rhodoviolascin (spirilloxanthin)	benzene	30	> 2	41	59 : 41	(364)
Ketones	Canthaxanthin (4,4'-diketo-β-carotene)	benzene	30	4	38	62 : 29 : 4 : 5	(123)
Hydroxyketones	Capsanthin	benzene	45	3	35	22 : 78	(503, 367)
	Capsorubin	benzene	15	2	20	20 : 80	(503)
Epoxides	β-Carotene monoepoxide	hexane	60	5	9	5 : 91 : 4	(440)
	β-Carotene diepoxide	hexane	60	1	2	98 : 2	(440)
	Taraxanthin (trihydroxy-α-carotene monoepoxide?)	benzene	30	3	17	17 : 83	(536)

Table 10. Composition of Stereoisomeric Mixtures Obtained from All-*trans*-β-carotene and its Main Isomers by 60-min. Refluxing of the Petroleum Ether Solution (b. p. 60–70°), in Weak, Diffuse Daylight (*365*).

Starting material	Relative photometric values in % of the recovered pigment			
	Neo U	All-*trans*	Neo B	Other *cis* forms
Neo-β-carotene U	31	40	19	10
All-*trans*-β-carotene	4	86	8	2
Neo-β-carotene B	4	50	40	6

Examples. (a) All-*trans*-β-carotene, at 190° for 15 min.; 33% of the recovered pigment remained unchanged and six *cis* forms were observed. (b) Neo-β-carotene U at 135° for 15 min.; ratio, unchanged neo U : all-*trans* : seven *cis* forms = 40 : 22 : 38. (c) All-*trans*-zeaxanthin, with 4 parts of naphthalene at 180° for 1 min.; unchanged all-*trans* : *cis* forms = 56 : 44.

Analogous rearrangements were realized by heating *cis* dimethylcrocetins (p. 114) in KBr discs above their melting points. After cooling, the IR spectrum of the all-*trans* compound was observed (*263*).

Stereoisomerization in the fused or sintered state is responsible for the phenomenon of double melting points (p. 19).

2. Photochemical *cis-trans* Isomerization in the Absence of Catalysts.

Each carotenoid solution studied so far was found to be more or less photosensitive, depending on the structure and configuration. As expected the wavelengths corresponding to the fundamental band are most effective.

In our laboratory the solutions are illuminated either by artificial light (Photoflood bulbs or Mazda fluorescent lamps) or exposed to intense sunlight ("insolation"). When necessary, they are protected by an inert gas and/or cooling.

As a rule photo-stereoisomerization competes with irreversible side-reactions. In part only slight structural alterations occur, with preservation of the strong chromophore, but partial cleavage to colorless (fluorescent) substances may also take place. The destruction is promoted by the presence of air and accidental catalysts. On overexposure complete bleaching may be observed.

An all-*trans* carotenoid must loose some of its color intensity during illumination, while in the case of mono- or di-*cis* isomers either a partial bleaching effect or the bathochromic effect of the *cis* → *trans* rearrangement may prevail. When a poly*cis* compound is insolated for a short time, the tremendous increase in color intensity predominates. Thus, very dilute prolycopene solutions that appear almost colorless to the naked eye turn intensely yellow while exposed to sunshine for a few minutes (*492, 515*).

Examples. (a) Upon 45-min. insolation of petroleum ether solutions at 30° the following ratios of unchanged to stereoisomerized starting material were found in the recovered pigment. All-*trans*-β-carotene, 98 : 2; neo-β-carotene B, 5 : 95; neo-β-carotene U, 37 : 63 (*525*). (b) When a dilute benzene solution of all-*trans*-zeaxanthin was insolated for 15 min., $^1/_6$ of the color intensity was lost. Composition of the recovered pigment, 48% unchanged all-*trans* form, 11% *cis* zeaxanthins, 12% minor pigments, and 29% crystallizable new carotenoid (*512*). (c) A dilute hexane solution of *retro*-dehydrocarotene was illuminated with two Mazda lamps from 60 cm. distance for 8 hours. Irreversible loss, 6%; *cis* forms present in the recovered pigment, 11% (*537*).

The behavior of the totally-synthetic, photolabile central-mono*cis*-β-carotene (which does not occur in equilibria ex all-*trans*-β-carotene) was described by INHOFFEN et al. (*187*). In diffuse daylight, at room temperature, this isomer rearranged almost completely within an hour; the wavelengths corresponding to the main band were most effective. Since those above 550 mμ do not act, this isomer (as some other similar pigments) can be handled safely in red light. It is remarkable that all-*trans*-β-carotene was practically the sole product of this illumination. In contrast, the iodine-catalyzed equilibrium mixture, formed within 3 minutes, contained only half of the pigment in the all-*trans* configuration. The *cis* configuration remained unchanged during a 20 hours' stay in darkness, in the presence of iodine.

For photochemical stereoisomerizations in rigid solvents cf. p. 44.

3. *cis-trans* Isomerization by Iodine Catalysis, in Light.

The catalytic action of iodine on carotenoid solutions at room temperature, in light, is the most frequently used method of stereoisomerization and yields within minutes a quasi-equilibrium mixture which can be resolved chromatographically. The process being reversible, each member of the set (or a mixture of members) should afford, after elution and renewed catalysis, the same stereoisomeric mixture. This postulate turned out to be very helpful in practical experimentation:

(a) If the investigator is interested in the isolation of a certain (minor) *cis* form only, that zone is cut out and set aside, while the mixture of all other pigments is subjected to a second treatment with iodine. These operations may be repeated and the desired *cis* zones combined. (b) Any *cis* carotenoid, even a non-crystallizable minor isomer, can be estimated quantitatively as follows, provided that the corresponding all-*trans* pigment is available in crystalline form: After the *cis* isomer has been eluted from its adsorbate and treated with iodine, one determines the extinction at λ_{max} of the resulting stereoisomeric mixture. This value is then compared with the molecular extinction coefficient at λ_{max} of an

4*

iodine-catalyzed equilibrium mixture which was prepared from a weighed sample of the all-*trans* compound.

The mechanism of iodine-catalyzed photoisomerizations about ethylenic double bonds is not yet clearly understood and will not be discussed here in detail.

It has been postulated repeatedly that halogen is added and a free-radical formed at an intermediate stage of the catalytic process [cf. e. g., KHARASCH (*246*); URUSHIBARA (*441*)]. It should be noted in this connection, however, that iodine is only one of several paramagnetic substances (O_2, NO, NO_2, Br_2, Na, K, etc.) which are effective catalysts. HARMAN and EYRING (*150*) proposed that magnetic interactions are responsible for catalytically promoted *cis-trans* rearrangements. According to McCONNELL (*309*), when a doublet electronic state of a catalyst atom interacts with the singlet and triplet states of an isomer, two doublet states are formed whose minimum separation is determined by the strength of the chemical binding between catalyst and substance. It is assumed that even weak *chemical* interactions may account for the observed effect. These arguments are perhaps preferable to the older concept that iodine promotes *cis-trans* rearrangements by adding to the molecule and then "breaking" the double bond.

The history of iodine-catalyzed rearrangements about ethylenic double bonds can be traced back to ANSCHÜTZ (*15*) (1878) who obtained ethyl fumarate by the interaction of silver maleate and ethyl iodide, because his ethyl iodide was contaminated with free iodine; the latter rearranged catalytically the primary reaction product, viz. ethyl maleate. In 1929 KARRER et al. (*230*) were able to convert, by means of iodine, the natural product bixin into "stable bixin"; and they interpreted the reaction tentatively as a geometrical isomerization (p. 107). This was proved to be correct by KUHN and WINTERSTEIN (*270*).

In the field of the C_{40}-carotenoids iodine catalysis was introduced by TUZSON and the writer (*536*) as late as 1939, closely followed by similar studies in collaboration with CHOLNOKY and POLGÁR (*503, 504*). Stress was laid on the reversibility of the process — a feature of practical importance that had not been claimed earlier. It was also demonstrated that in general iodine does not rearrange C_{40}-carotenoids in darkness (Fig. 7, p. 31). (For the behavior of all-*trans* and *cis* forms of compounds with shorter chromophores cf. p. 131.)

In our early experiments, conducted in diffuse laboratory daylight, the concentration of the pigment was roughly 0.1 mg. per ml. of hexane or benzene ($^1/_{5000}$ molar) and the weight of the iodine amounted to 1–2% of the pigment. Stereoisomeric quasi-equilibrium was reached within 15–60 min. but in some instances within a few minutes.

Example. Lycopene in benzene solution (*536*):

Minutes of catalysis	0	5	30	90	180
Ratio, all-*trans* : *cis* forms	100 : 0	82 : 18	71 : 29	57 : 43	53 : 47

Table 11. Examples of All-*trans*→*cis* Isomerizations of Carotenoids, Catalyzed by Iodine, in Light.

(In the column "Ratio of steric forms" the first figure designates the top zone on the chromatographic column; data referring to unchanged all-*trans* forms are italicized; trace amounts are neglected.)

Type of compound	Name	Observed number of *cis* forms	Extent of isomerization in the recovered pigment (%)	Ratio of steric forms in the recovered pigment	References
Hydro-carbons	Lycopene..........	2	43	57:43	(536, 515)
	α-Carotene	7	48	14:3:15:52:13:3	(525, 332)
	β-Carotene..........	12	52	22:48:25:3:2	(365)
	γ-Carotene..........	>5	47	3:53:7:14:16:7	(526)
	5,6-Dihydro-α-carotene........	5	44	4:20:56:16:4	(366)
	5,6-Dihydro-β-carotene	5	33	18:67:15	(366)
	3,4-Dehydro-α-carotene........	4	32	7:2:68:6:17	(221)
	3,4-Dehydro-β-carotene	4	51	22:49:26:3	(221)
	retro-Dehydrocarotene	>8	80	20:32:5:12:14:5:7: 4:1	(537)
	Neurosporene	6	56	44:13:11:12:15:3:2	(300)
	Phytofluene	1	10	90:10	(350, 250)
Alcohols	Cryptoxanthin (3-hydroxy-β-carotene)	3	41	18:59:18:5	(504, 512)
	4-Hydroxy-α-carotene	6	37	13:63:21:3	(48)
	Zeaxanthin (3,3'-di-hydroxy-β-carotene)	3	34	10:21:3:66	(536, 512)
	Lutein (3,3'-di-hydroxy-α-carotene)	2	40	17:23:60	(536)
	5,6-Dihydroxy-5,6-di-hydrolycopene	4	49	11:51:12:17:9	(48)
Ethers	4-Ethoxy-α-carotene .	5	39	13:1:61:11:8:6	(48)
	Rhodoviolascin (spirilloxanthin)	2	40	60:40	(364, 411)
Esters	Physaliene (zea-xanthin dipalmitate)	1	55	45:55	(504)
	Capsanthin dipalmitate	2	33	67:33	(503)
	Capsorubin dipalmitate........	2	25	75:25	(503)
Mono-ketones	4-Keto-α-carotene ...	4	48	11:52:19:12:6	(48)
Diketones	Canthaxanthin (4,4'-diketo-β-carotene) ..	6	38	62:18:10:5:5	(123)

(continued on next page)

(Table 11, continued.)

Type of compound	Name	Observed number of *cis* forms	Extent of isomerization in the recovered pigment (%)	Ratio of steric forms in the recovered pigment	References
Hydroxy-ketones	Capsanthin	3	31	15:11:5:69	*(503, 367)*
	Capsorubin	2	32	32:68	*(503)*
Epoxides	β-Carotene mono-epoxide	9	44	14:1:15:56:7:5:2	*(440)*
	β-Carotene diepoxide .	7	41	4:25:59:10:2	*(440)*
	Taraxanthin (tri-hydroxy-α-carotene monoepoxide?)	3	44	44:56	*(536, 108)*

It was recognized later that brief illumination by means of standardized artificial light is the method of choice: One or several well-filled and stoppered 25- or 50-ml. Pyrex volumetric flasks, each containing a hexane or benzene solution of 5–25 mg. of pigment and 1–2% iodine (pigment = 100%), were exposed for 5–45 min. to the light of two parallel Mazda fluorescent lamps (40 W, 3500°, white and yellow) from 60 cm. distance.

Table 11 above shows that the resulting stereoisomeric mixtures contained $^1/_3$–$^1/_2$ of the pigment in *cis* configurations, with a few exceptions for which some special structural features were responsible.

The stereoisomeric phytofluene mixture contained only 10% *cis* forms *(250)*. In contrast, *retro*-dehydrocarotene isomerized to the extent of 80% *(537)*. Indeed, in this instance the amount of one main mono*cis* isomer by far surpassed the unchanged portion of the all-*trans* compound. This feature is in accordance with the finding that the preparation of *retro*-dehydrocarotene, when carried out by the interaction of β-carotene and iodine, followed by a treatment with thiosulfate *(266)*, leads mainly to *cis* forms, contrasting with most structural conversions of *normal* carotenoids *(537)*.

The main factors that influence iodine-catalyzed rearrangements are, the concentrations of pigment and catalyst, the absolute quantity of the iodine, the pigment/iodine ratio, and the mode and duration of the illumination. In light, iodine does not act exclusively as a promoter of *cis-trans* rearrangements but also participates in irreversible side-reactions. Thus, some pigment may be lost, especially on over-exposure.

For α-carotene ZSCHEILE et al. *(542)* have estimated the irreversible losses. Further, they found that the effective radiation is that absorbed by the iodine rather than by the carotene, and that in prolonged experiments the free iodine gradually disappears from the solution. Of course, catalysis by iodine cannot succeed if either the solvent or the plant extract contains unsaturated impurities which add halogen at high rates and thus remove the catalyst from the system *(542)*. A single chromatographic operation will usually correct such situations.

STRAIN (*418*) has reported that, under certain conditions, zeaxanthin yielded (in part) the same irreversible pigments with halogen which were obtained by means of acids (see below). The presence of a little organic base prevented this from happening.

The nature of a few irreversible side-reactions has been clarified: (a) 4-Hydroxy-α-carotene afforded small amounts of the corresponding ketone (*48, 49*). (b) Either of the two β-carotene epoxides was converted to a limited extent into furanoid oxides (*440*). In the case of another epoxide, trollixanthin, no iodine-catalyzed *cis* → *trans* isomerization could be realized because of conversion into a furanoid oxide (*108*).

In spite of the possible complications that were rather emphasized above, iodine, when applied under proper conditions, is a very efficient tool in stereochemical research.

When working with carotenoid pigments it should also be considered that the degree of iodine sensitivity is a function of the number and type of *cis* double bonds. Especially sensitive are sterically hindered such bonds; these afford the equilibrium mixture at much higher rates than unhindered *cis* or *trans* double bonds. Very similar is the behavior of the naturally occurring poly*cis* compounds which disappear rapidly from iodine-containing solutions (pp. 68, 7ɔ).

Iodine sensitivity and degraded spectrum go together in many instances. Although this statement is also valid in the diphenylhexatriene and diphenyloctatetraene sets (*292, 522*) (pp. 172, 177) it does not apply to dienes. Certain isoprenic, hindered *cis* dienes of the C_{20}-series are resistant to iodine while some unhindered *cis* isomers respond to the catalyst normally (*339*). In two polyene-azine sets (pp. 172, 20ɔ) *cis* C=C bonds were found to be more sensitive to iodine than C=N bonds but no pertinent generalization is yet possible (*84*).

It would be well worth while to carry out systematic kinetic studies in the field just discussed.

4. *cis-trans* Isomerization by Acid Catalysis.

Although this method may be useful in certain instances, its application has been abandoned in our laboratory, because the spatial rearrangement was frequently accompanied or overruled by irreversible processes. It has been recognized long time ago that especially the xanthophylls are acid-sensitive, even in the presence of weak acids (*480, 273, 416, 418, 424, 423, 377*). A classical example of acid sensitivity is the behavior of KARRER's carotenoid epoxides: Even the acid traces present in commercial chloroform convert them quantitatively into furanoid oxides, at room temperature, within a few seconds [KARRER (*222*)]. A similar H^+-catalysis causes the almost instantaneous elimination of allylic hydroxyl groups.

Example. 4-Hydroxy-α-carotene→3,4-dehydro-α-carotene [BUSH et al. (*49*)].

Carotenes are much less acid-sensitive. When a petroleum ether solution of β-carotene was shaken with strong hydrochloric acid, about half of the pigment underwent stereoisomerization but very little was changed irreversibly. Application of strong hydriodic acid afforded 5,6-dihydro-β-carotene [POLGÁR et al. (*366*)].

5. *cis-trans* Isomerization by Contact with Active Surfaces.

As mentioned in connection with GILLAM's pioneer experiments (p. 8), the all-*trans* configuration of carotenoids is not affected by routine chromatography. According to MEUNIER et al. (*313*), however, this situation may be altered under special conditions. The conditions applied by these authors were static and hence essentially different from those existing in the chromatographic system: Very dilute pigment solutions, contained in conical tubes, were left in contact (without shaking) with a large excess of strongly active Al_2O_3 or TiO_2, for days and weeks, protected from light and air. A treatment of β-carotene resulted in the appearance of a marked *cis*-peak observed in the supernatant petroleum ether solution; and the intensity of the peak seemed to indicate a *trans* → *cis* shift in the middle section of the molecule. No spectral change was noticed in the absence of the adsorbent within a month. MnO_2 also caused the described spectral effect but this was soon overruled by the progressing cleavage of carotene to retinene (*312*).

The stereoisomerization of fucoxanthin is accelerated by contact with sugar or glass powder (p. 103).

Evidently, a more profound study into the behavior of carotenoids on strongly active surfaces would be desirable.

A reported stereoisomerization of astaxanthin on alumina needs confirmation, since the pigment was also exposed to acetic acid (*142*). An attempt to stereo-isomerize β-carotene epoxides by MEUNIER's method did not afford *cis* compounds but furanoid oxides (*440*). For the influence of moist Celites on carotenoids cf. (*482*).

6. *cis-trans* Isomerization via Boron Trifluoride Complexes.

PRICE and MEISTER (*374*) predicted long ago that this reagent, when added to ethylenic double bonds, will afford on cleavage *cis-trans* equilibria; an alleged application of this principle to stilbene, however, turned out to be erroneous (*373, 374*). Our recent experiments have shown that BF_3 etherate may well become a tool for the preparation of certain *cis* carotenes, under strictly defined conditions (*476, 477, 49, 499*).

Example. After a 1-min. interaction of BF_3 and β-carotene the mixture was composed of 36% unchanged all-*trans* form, 44% *cis* isomers, 1% foreign carotenoids, and 19% unidentified (in part colorless, fluorescent) substances. In the case of *retro*-dehydrocarotene profound structural changes took place within a minute.

7. Bio-stereoisomerization.

It was interesting to learn that ingested carotenoids may alter their configurations in the body. The factors directing such rearrangements are more complex than those dealt with in the above Sections. They may include thermal and pH effects, the action of unknown catalysts, the metabolism of intestinal bacteria, etc. Some observations available in this field will be mentioned in Chapters XI and XII (pp. 118, 121).

VI. Preparation of *cis* Carotenoids by Total Synthesis.

The most brilliant achievement in polyene chemistry during the last decade has been the total synthesis of certain naturally occurring carotenoids and of some analogs and homologs that have not yet been found in plants. The synthesis of β-carotene was announced in 1950 by KARRER and EUGSTER (226), almost simultaneously by INHOFFEN, BOHLMANN, BARTRAM and POMMER (184), and somewhat later by MILAS et al. (316). Since this breakthrough the new field has been broadened and deepened by the research groups of KARRER, INHOFFEN, ISLER, WEEDON, and others [cf. also SURMATIS et al. (428)]. The pertinent literature till 1952 has been discussed in detail by INHOFFEN and SIEMER (193); for a brief survey cf. KARRER (223). In 1957 ISLER and ZELLER's excellent review has appeared (210). The present report, which of course cannot do justice to the authors active in this fascinating field, will be restricted to a few stereochemically important aspects of carotenoid synthesis.

The C_{40}-molecule was built up in most instances by joining three fragments, according to the schemes, $C_{10} + C_{20} + C_{10}$; $C_{14} + C_{12} + C_{14}$; $C_{15} + C_{10} + C_{15}$; $C_{16} + C_8 + C_{16}$; $C_{18} + C_4 + C_{18}$; or $C_{19} + C_2 + C_{19}$. In the last two cases diacetylene or acetylene units were incorporated into the chain by means of Grignard derivatives, lithium-organic compounds, etc.

As characteristic intermediates, C_{40}-glycols were thus obtained that contain ethylenic and acetylenic bonds but are void of a complete conjugated system. Hence, at this stage the two following transformations were required: (a) elimination of the OH-groups by dehydration with formation of double bonds; and (b) partial reduction of the triple bond(s) to double bond(s). In some of the syntheses step (a), in others step (b) was carried out first *(Charts 1 and 2)*. When (a) preceeds (b), the reaction becomes stereochemically important because it is possible to conduct the partial saturation of the acetylenic bond(s) in a stereospecific manner (344, 8, 114; surveys, 73, 47). A number of *cis* carotenoids have thus been obtained in which the location of the *cis* double bond(s) is established beyond doubt *(Table 12)*.

In some instances the synthesis has yielded three intermediate pigments that contain, respectively, two triple bonds, two cumulene groups, and/or a triple bond and a cumulene group (besides other olefinic bonds). After chromatography, each of these compounds was submitted to partial saturation [GARBERS, EUGSTER and KARRER (*125*); cf. KUHN and FISCHER (*262*).]

A few remarks on the partial reduction of triple bonds follow. SLY's X-ray crystallographic study (*407*) of a typical intermediate of the β-carotene synthesis, to wit the central-acetylenic compound 15,15'-dehydro-β-carotene (formula XXXIII, p. 61) has shown that this molecule possesses axial symmetry and hence a straight over-all shape in the crystal. The situation is, however, different when dissolved molecules are in contact with an appropriate catalyst. Then this compound may react either in its straight form or in a form bent at the center *(Fig. 17*, top, p. 62) (*185*). Consequently, the configuration about the emerging double bond will depend on experimental conditions; for instance, LiAlH$_4$ affords the all-*trans*, but the Lindlar catalyst a *cis* form (*189*).

Table 12. Examples of Totally-synthetic *cis* Carotenoids.

Stereoisomeric set	Configuration	References
β-Carotene	central-mono*cis*	(*185, 187, 203, 485*)
	11,11'-di*cis* (hindered)	(*107, 195*)
	cis-β-carotene C (hindered) . . .	(*107, 195*)
3,4-Dehydro-β-carotene	central-mono*cis*	(*204*)
3,4,3',4'-Bisdehydro-β-carotene	central-mono*cis*	(*204, 191*)
13,13'-Bis-desmethyl-β-carotene C$_{38}$H$_{52}$	central-mono*cis*	(*186*)
16,16'-*Homo*-β-carotene C$_{42}$H$_{58}$	middle-mono*cis* (16-*cis*)	(*183*)
	middle di*cis* (16,16'-di*cis*)	(*183*)
α-Carotene	central-mono*cis*	(*192*)
Lycopene	central-mono*cis*	(*197*)
	hindered *cis* forms b, c and c'	(*128*)
Cryptoxanthin	central-mono*cis*	(*202*)
Zeaxanthin	central-mono*cis*	(*201, 207*)
Isozeaxanthin	central-mono*cis*	(*205*)
Echinenone	central-mono*cis*	(*5*)
Canthaxanthin	central-mono*cis*	(*205, 206, 539*)
Torularhodin methylester	central-mono*cis*	(*196*)
Apo-β-carotenals	15,15'-*cis*	(*393, 394*)
1,18-Diphenyl-3,7,12,16-tetramethyl-octadecanonaene	5-*cis* (hindered)	(*106, 125, 126*)
	5,13-di*cis* (hindered)	(*106, 125, 126*)
	other hindered and unhindered *cis* isomers	
1,3,7,12,16,18-Hexaphenyl-octadecanonaene	four hindered *cis* forms and from them (with iodine) several unhindered ones . . .	(*540*)
"Naphthyl-carotene"	central(?)-mono*cis*	(*287*)

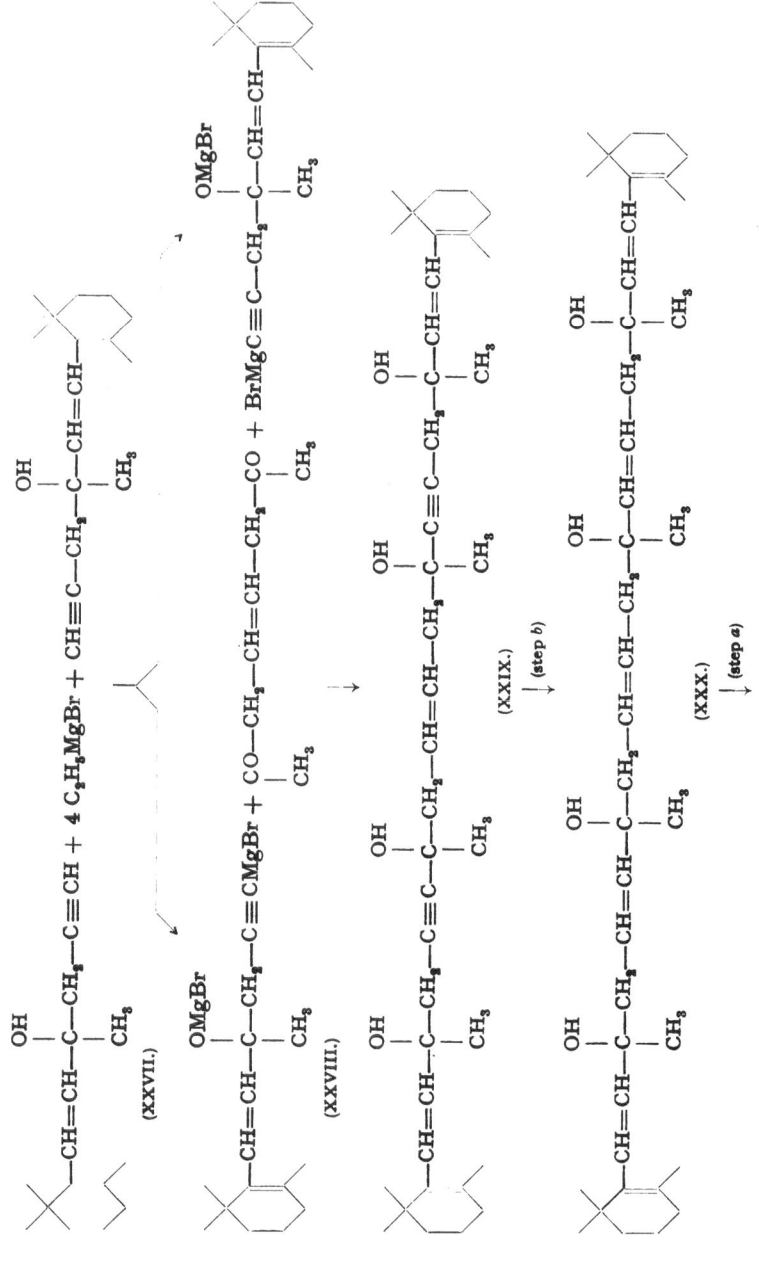

Chart 1. Synthesis of β-Carotene according to KARRER and EUGSTER (*227*).

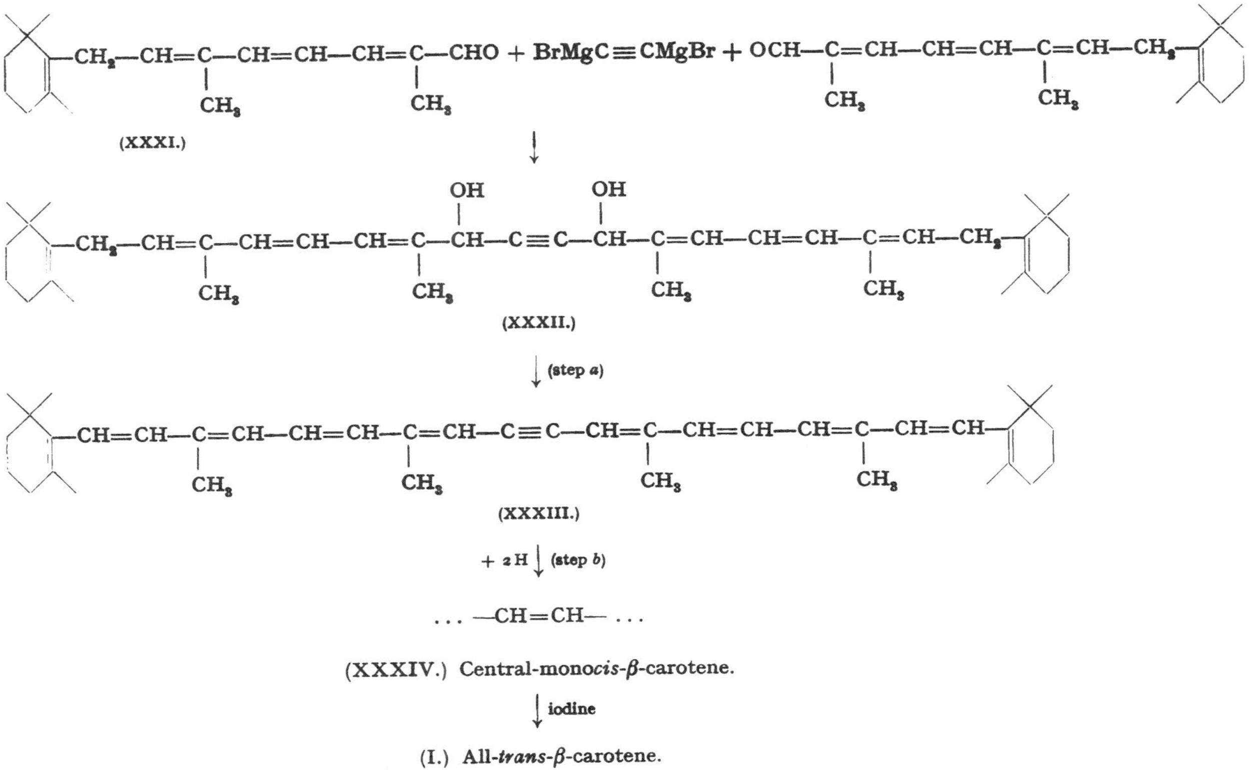

$$CH_3-CH=C-CH=CH-CH=C-CHO + BrMgC \equiv CMgBr + OCH-C=CH-CH=CH-C=CH-CH_3$$

(XXXI.)

$$\downarrow$$

$$CH_3-CH=C-CH=CH-CH=C-CH=C-CH \equiv C-CH-C=CH-CH=CH-C=CH-CH_3$$

(XXXII.)

$$\downarrow \text{(step 4)}$$

$$CH=CH-C=CH-CH=CH-C=C-CH=C-CH=CH-CH=C-CH=CH$$

(XXXIII.)

$$+ 2 H \downarrow \text{(step b)}$$

$$\ldots \; -CH=CH- \ldots$$

(XXXIV.) Central-mono*cis*-β-carotene.

$$\downarrow \text{iodine}$$

(I.) All-*trans*-β-carotene.

Chart 2. Synthesis of β-Carotene according to INHOFFEN et al. (*185*) (see also Models IV and I on p. 14).

The well-known Lindlar catalyst is prepared by deactivating Pd or Pd/CaCO₃ by means of hot lead acetate solution (*286*). In some instances the presence of quinoline secures maximum efficiency and/or specificity. The interaction with carotenoids is best carried out in darkness or red light considering the photosensitivity of some *cis* compounds.

Cumulenes can be reduced to *cis* polyenes by means of the same catalyst (*262, 125*), cf. p. 189.

A comparison of totally-synthetic *cis* carotenoids with our neo compounds has confirmed and extended numerous earlier observations concerning

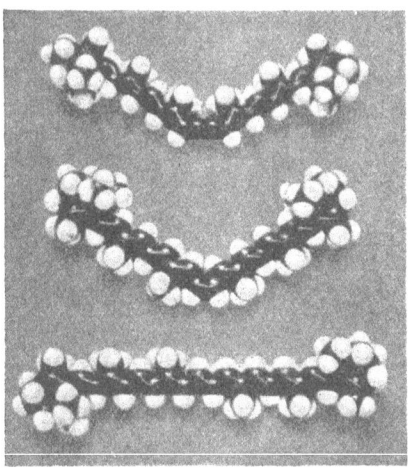

Fig. 17. Models of 15,15'-dehydro-β-carotene (top), central-mono*cis*-β-carotene (middle), and all-*trans*-β-carotene (bottom); according to INHOFFEN et al. (*185*). [From: Liebigs Ann. Chem. 570, 54 (1950).]

adsorption affinities, spectral curves, stabilities, λ_{max} shift, *cis*-peak effect, and so forth. The synthetic studies have also shown, however, the availability of hindered *cis* isomers that failed to appear in our stereoisomeric equilibria and whose very existence had been questioned at a certain stage of research (p. 16).

No synthetic poly*cis* carotenoid with known configuration has been reported up to the present time, and it would be desirable to carry out such theoretically important work in the near future.

VII. Naturally Occurring *cis* and Poly*cis* Carotenoids.

1. Mono- and Di*cis* Carotenoids.

As we have seen, the overwhelming majority of plant carotenoids possess the all-*trans* configuration. To what extent the more energy-rich *cis* carotenoids appear in plants and animals is a question of biological and genetical interest. The relative scarcity of *cis* carotenoids in living tissue indicates the presence of some directing and protective mechanism that forms and preserves the all-*trans* configuration, even when the pigment is exposed to strong sunlight and heat; similar conditions cause *trans* → *cis* rearrangements in extracts. After a *cis* carotenoid has been formed, its configuration is likewise protected in the plant.

Examples. (a) Large quantities of *cis*-phytofluene do occur in the tomato but upon extraction rapid isomerization to the all-*trans* form takes place, especially in light (*350*, *250*). (b) Although intact red pepper fruits contain (almost) exclusively all-*trans* pigments, exposure to sunshine of the opened fruit results in stereo-isomerization (*305*, *306*). (c) It has been shown by SAPERSTEIN and STARR (*396*) that some carotenoids of non-photosynthetic bacteria do not undergo *trans* → *cis* rearrangements under the influence of iodine and light, while attached to protein. (This should be further confirmed by showing that the catalyst is not used up by the complex.)

In regard to the natural occurrence of neo carotenoids that contain one or two *cis* double bonds, the literature appears to be in a state of confusion. Because of the spontaneous attainment of *trans-cis* equilibria, even at room temperature, a number of *cis* isomers claimed as natural products, might well have been formed during extraction, purification, storage, evaporation at elevated temperature, exposure to alkali, etc. Whether or not certain chromatographic *cis* zones contained "genuine" pigment remains open to criticism.

Although some authors have accepted the observed "occurrence" of *cis* forms at face value, others did recognize the situation correctly. According to PORTER and ZSCHEILE (*370*), the neo-β-carotene U found in canned tomatoes resulted from isomerization during the canning process. STRAIN (*416*) emphasized the necessity of extracting carotenoids under mild conditions. COOK (*70*) when reviewing algal pigments, has expressed doubts about the occurrence of some of the reported isomers. Similar statements are valid for some pigments in petals (*61*), in *Rhodospirillum* (*215*), in *Haematococcus* (*139*), in *Penicillium* (*307*), in fossil sediments (*443*), etc.

It should also be remembered that previous storage conditions of plant materials may be responsible for the occurrence of some *cis* carotenoids in extracts. Thus, high temperatures applied during dehydration of alfalfa meal promotes *trans* → *cis* changes; and exposure of the meal to visible light decreases the neo-β-carotene B content while more neo U appears [BICKOFF et al. (437, 414)].

The situation is much clearer when the observed *cis* carotenoid does not occur in any detectable amounts in stereoisomeric equilibria. Such an isomer would disappear rapidly upon the addition of iodine to the extract and illumination. This happens, for example, in the case of the poly*cis* compounds and of neo-γ-carotene P, a constituent of *Pyracantha* berries (521).

Likewise, a clear decision can be made when a native pigment, which, although a component of the equilibrium, occurs in the plant either as the sole or as the preponderant member of the set. For example, according to TAPPI and KARRER (432), the main pigment of the pollen sacs of *Lilium candidum* is a crystallizable mono*cis*-antheraxanthin (zeaxanthin monoepoxide) which also occurs in *L. umbellatum, L. regale, L. Willmottiae unicolor, L. Maxwill*, and *L. mantchuricum* (229, 433). (The corresponding all-*trans* compound was isolated from some other species.) Along slightly different lines, SUZUKI and TSUKIDA (429) have shown that neo-β-carotene B is a genuine constituent of *Osmanthus fragrans* flowers, because the fresh extract was free of neo-β-carotene U which is contained in amounts matching those of neo B at equilibrium.

When the interconversion rates of the stereoisomers are very low, then the probability of their natural occurrence increases. This may be assumed for the three stereoisomeric fucoxanthins, "unless a rapid conversion takes place immediately upon death of the cell" (424, 426).

Under less favorable circumstances it may still be possible to decide, as follows, whether or not an observed *cis* carotenoid is a natural product.

Two equal samples of the plant material are taken and one of them is mixed with a solution of the all-*trans* pigment, in an amount similar to that present in the natural material. All extraction and purification operations are then carried out with both samples simultaneously and the two final chromatograms are compared. If the column originating from the sample to which pigment had been added does not show an increase of the *cis* zones, a strong argument has been gained in favor of natural occurrence.

2. Poly*cis* Carotenoids.

The first representative of this class was detected as late as 1941 in some tomato varieties such as the "tangerine tomato". The reason that had prevented its earlier detection is its absence from stereoisomeric equilibria ex all-*trans*-lycopene. The color of the ripe tangerine tomato is orange instead of red. Because the pigment substituted for ordinary lycopene, it was termed "prolycopene" (516). A few similar pigments have been isolated since. Most vegetable tissues are free of poly*cis* carotenoids, quite a number of plants contain trace amounts, but considerable quantities have rarely been found. More than 20 mg. of crystalline prolycopene was isolated from 1 kg. of fresh tangerine tomatoes (yield, 93%) [LE ROSEN, WENT, PAULING and the writer (*516*, *282*)].

A systematic search for naturally occurring poly*cis* pigments is still lacking. It will be noted that the relatively few higher plants listed in *Table 13* (p. 66) belong to ten families, and that poly*cis* pigments occur also in microorganisms. Although they are evidently wide-spread, poly*cis* carotenoids are not mass products of biosynthesis, in sharp contrast to all-*trans* forms.

With respect to the poly*cis* configuration of their carotenoids, even closely related plants may show considerable differences.

Example. The ripe berries of *Pyracantha angustifolia* harvested in California are rich in poly*cis* forms, but no such pigments have been found in *P. yunanensis* (California) or in *P. coccinia* (Europe) (*236*); the latter plants produce all-*trans*-lycopene.

Although almost all known poly*cis* carotenoids are hydrocarbons (Table 13), poly*cis* α- or β-carotenes are unknown.

As we have seen, the conspicuous characteristics of poly*cis* carotenoids are the unusually weak adsorption affinity, the degraded fundamental band and the very high value of the λ_{max} shift. No carotenoid should be assigned, however, a poly*cis* configuration on this basis alone; rather the impressive spectral shift on iodine catalysis should be considered as decisive (Figs. 18 and 20, pp. 68, 70).

Very little is known about the biosynthesis of poly*cis* carotenoids; the suggestion that prolycopene may be a normal precursor of all-*trans*-lycopene in tomatoes does not seem to hold (*137*, *212*, *369*, *438*).

WENT and the writer (*538*) have considered this problem in the light of the chemical steering power of genes which is generally directed toward a single structural target. In some instances, we believe, a specially high degree of selectivity is required in order to build up a certain steric configuration. It is expected that, as in the case of the tomato, in many other instances one gene will be responsible for the synthesis of the polyene structure but another gene will steer the molecules to a predetermined configuration. This latter function may be important for the plant, because *cis-trans* configurations determine the overall molecular shape and because only certain geometrical forms may fit into a given biological system.

In general, the stereochemical status of plant polyenes, poly*cis* carotenoids included, is strongly influenced by both genetical and environmental factors.

Along genetic lines, by means of crossing and selection experiments, detailed studies of the tomato fruit have been reported by PORTER and ZSCHEILE (*370*),

Table 13. Examples of the Occurrence of Polycis Carotenoids in Nature.

Compound	Plant family	Plant	Organ	References
Prclycopene*	Solanaceae	*Lycopersicum esculentum* (MILL.) var. "Tangerine tomato"	Fruit	(516, 282, 521)
		var. "Golden Jubilee"	Fruit	(370, 369)
		L. esculentum, L. hirsutum, L. pimpinellifolium L. (crossings, selections)	Fruit	(298)
	Pomoideae	*Pyracantha (Cotoneaster) angustifolia* SCHNEID.	Fruit	(530)
	Celastraceae	*Evonymus fortunei* REHD. (Winter creeper)	Seed hulls	(508)
	Scrophulariaceae	*Mimulus longiflorus* GRANT (Monkey flowers)	Flowers	(398)
	Palmae	*Butia capitata* BECC., *B. eriospatha* BECC.	Fruit	(528, 531)
			Fruit	(17)
Polycis-lycopenes I–VI	Algae	*Chlorella vulgaris* mutant	—	(63–65)
Polycis-lycopene I	Pomoideae	*Pyracantha angustifolia* SCHNEID.	Fruit	(521)
Polycis-lycopenes	Compositae	*Calendula officinalis* L. (Pot marigold)	Flowers	(136)
	Araceae	*Arum maculatum* L. (Cuckoo pint)	Fruit	(138)
	Rosoideae	*Rosa canina* L., *R. moyesii* (Rose hips)	Fruit	(138)
	Cruciferae	*Brassica rutabaga, B. napus* L.	Root	(218, 219)
Polycis-lycopenes (seven)	Solanaceae	*Lycopersicum esculentum* L. (strains, crosses, selections)	Fruit	(297, 369, 439, 300, 370)
Proneurosporene (protetra-hydrolycopene, neoneurosporene P, polycis-ψ-carotene, unidentified carotene I)	Palmae	*Elacis guineensis* JACQ.	Fruit	(17)
	Algae	*Chlorella vulgaris* mutant	—	(63–65)
	Palmae	*Butia capitata* BECC., *B. eriospatha* BECC.	Fruit	(528, 531)
Pro-γ-carotene	Palmae	*Elacis guineensis* JACQ.	Fruit	(17)
	Pomoideae	*Pyracantha angustifolia* SCHNEID.	Fruit	(521, 530)
	Celastraceae	*Evonymus fortunei* REHD.	Seed hulls	(503)
	Scrophulariaceae	*Mimulus longiflorus* GRANT	Flowers	(398)
	Sulfur bacteria	*Chlorobium spp.*	—	(140)
Monohydroxy-pro-γ-carotene	Pomoideae	*Pyracantha angustifolia* SCHNEID.	Fruit	(530)
Polycis-cryptoxanthin	Aurantioideae	*Citrus aurantium* RISSO (Valencia orange)	Fruit juice	(77)

* In our nomenclature "prolycopene" designates an individual polycis-lycopene, viz. the main pigment of the Tangerine tomato; we characterize some similar isomers as polycis-lycopenes I, ...

ZSCHEILE and PORTER (*543*), MACKINNEY and JENKINS (*297, 298*), PORTER and LINCOLN (*369*), TROMBLY and PORTER (*439*), as well as by MANUNTA (*304*); see also (*515, 212, 438*). Surveys on connected subjects have been pre.en.ed by MACKIN\EY (*296*) and by GOODWIN (*137*).

Turning now to environmental factors, light should be mentioned as of primary importance. This was shown by KUHN and WINTER-STEIN (*271*) in some experiments concerning crocetin (pp. 11 {-115).

In our laboratory, SCHROEDER (*398*) made the interesting observation that when cut stems (with buds on) of "Monkey flowers" *(Mimulus longiflorus)* were placed in water and the flowers were allowed to open in diffuse room light, the petals produced substantial amounts of the poly*cis* pigments prolycopene and pro-γ-carotene. Conversely, *cis* forms were missing from flowers developed on the bush, in the open, and the extracts contained only the two corresponding all-*trans* carotenoids. Furthermore, when the flowering plant was exposed to weak sunshine only, the petals remained yellow but during a period of prolonged clear weather the usual orange coloration did appear. Such observations also demonstrate how profoundly *cis-trans* isomerism may enrich the variety of colors in petals and fruits.

Possibly, similar environmental factors were responsible for the isolation of either *cis*- or *trans*-trollixanthin from the petals of *Trollius europaeus* [EUGSTER and KARRER (*108*)].

In the class of the green algae CLAES (*63, 64*) has reported that when the *Chlorella* mutant "5/520" (obtained by X-ray irradiation) was cultivated in darkness, it produced poly*cis* carotenoids (prolycopene and protetrahydrolycopene). However, illumination of the dark cultures with blue light resulted in the decrease and disappearance of the poly*cis* isomers and in the formation of all-*trans*-lycopene and all-*trans*-tetrahydro-lycopene instead. According to CLAES and NAKAYAMA (*65*) the photo-isomerization, poly*cis* → all-*trans*, in the algal cell can be realized even in red light, provided that chlorophyll is present. The same effect (which is inhibited by oxygen) can be demonstrated in vitro, with petroleum ether solutions of prolycopene or protetrahydrolycopene. It would seem that, under certain conditions, the spatial configuration of chloro-plast carotenoids might be controlled by chlorophyll (cf. also *64a*).

In the class of the Fungi RAU and ZEHENDER (*387*) found that *Fusarium aquaeductum* LAGH. develops poly*cis*-lycopene II when strongly ventilated in light.

Prolycopene.

Pure, crystalline prolycopene was described by LEROSEN and the writer (*282*). It is sharply differentiated from neo-lycopenes by the thermostability of solutions and easy crystallizability. Its melting point,

5*

111°, lies 64° below that of lycopene. When refluxed, prolycopene is at least as thermostable as all-*trans*-lycopene.

An impressive difference from the all-*trans* compound is in the color.

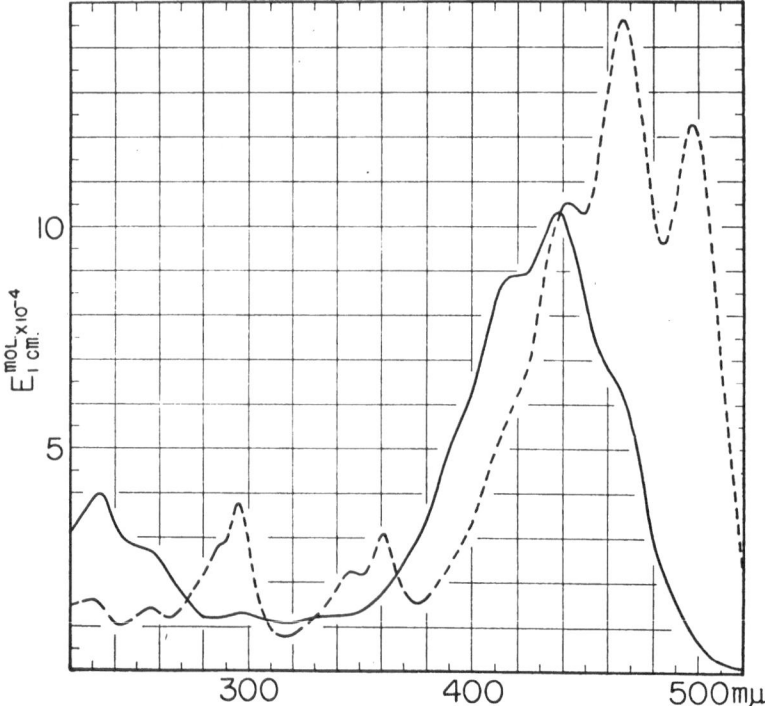

Fig. 18. Molecular extinction curves of prolycopene, in hexane: ————, fresh solution; and — — —, mixture of stereoisomers after catalysis by iodine (*521*). [From: J. Amer. Chem. Soc. 69, 1930 (1947).]

Under the microscope the crossing sections of individual prolycopene crystals are dull-brown, in contrast to the corresponding ruby-red color of all-*trans*-lycopene. When measured in a simple colorimeter, the color intensity of prolycopene in hexane is about one half of that of lycopene.

The spectral maximum of prolycopene in hexane solution is located at 438 mμ, i. e. at about 35 mμ shorter wavelengths than that of lycopene. The degraded curve shows much depressed extinction values and only little fine structure *(Fig. 18)*. A *cis*-peak is absent and hence the molecule must have an overall straight form.

Prolycopene is highly photosensitive in the presence of iodine. Upon addition of the catalyst and exposure to light a striking change takes place almost instantaneously and the intense spectral bands of the well-known lycopene-neolycopenes equilibrium mixture appear (Fig. 18).

Demonstration in the visual spectroscope: A hexane solution of prolycopene shows a blurred spectrum without distinct bands. Upon the addition of a few drops of a dilute hexane solution of iodine sharp bands appear within a minute; the light necessary for this rearrangement is supplied by the instrument. The same effect may be demonstrated simply by exposing prolycopene solutions in two small Petri dishes to the light of an ordinary electric bulb and adding a drop of iodine to one of the samples: the pale yellow color rapidly turns a deep orange. .

With reference to the configuration of prolycopene the following few comments can be made: The degraded character of the fundamental band could be explained a priori either by the presence of several unhindered *cis* double bonds or by that of at least one sterically hindered such bond (or both). When discussing similarly degraded fundamental bands of some totally-synthetic, hindered mono- and di*cis* polyenes, GARBERS, EUGSTER and KARRER (*128*, *126*, cf. *106*, *125*) seem to have assumed that the molecules of the natural poly*cis* carotenoids possess one or perhaps two hindered *cis* double bonds rather than a larger number of unhindered such bonds. However, as has been pointed out repeatedly [DALE (*79*), LUNDE et al. (*293*)], representatives of the two series can be distinguished in the *cis*-peak region where the curves of the natural poly*cis* carotenoids are flat while GARBERS and KARRER's substances (hindered *cis* lycopenes, β-carotenes, and 1,18-diphenyl-3,7,12,16-tetramethyl-octadecanonaenes) show marked *cis*-peaks (Figs. 37, 38, 28 and 42, pp. 99, 83, 104). This feature has not been considered by KARRER's research group. In the lycopene set a further difference lies in the high thermostability and easy crystallization of prolycopene while the synthetic, hindered *cis* lycopenes b, c and c' have not crystallized (*128*).

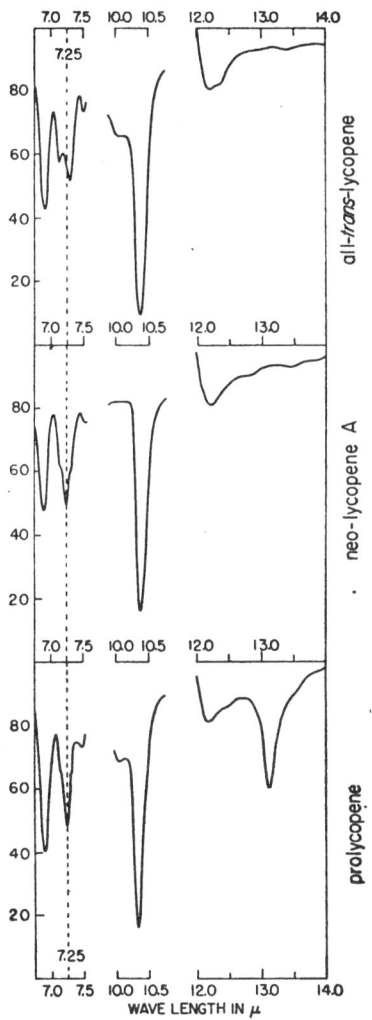

Fig. 19. Stereochemically important sections taken from infrared curves of some *cis-trans* isomeric lycopenes: 1% CCl₄ solutions (1 mm. cell) in the 7.0—7.5 μ and 10.0—10.5 μ regions; 0.25% cyclohexane solutions (2 mm. cell) in the 12.0—14.0 μ region (*293*). [From: J. Amer. Chem. Soc. 77, 1647 (1955).]

Another approach to the stereochemistry of prolycopene, viz. stepwise rearrangement was initiated some time ago (*282*). A prolycopene solution was treated with unusually small amounts of iodine and the resulting mixture chromatographed. The column showed several pigment zones located between those of unchanged prolycopene and the very little all-*trans*-lycopene formed. A hindered mono*cis* isomer should have yielded the

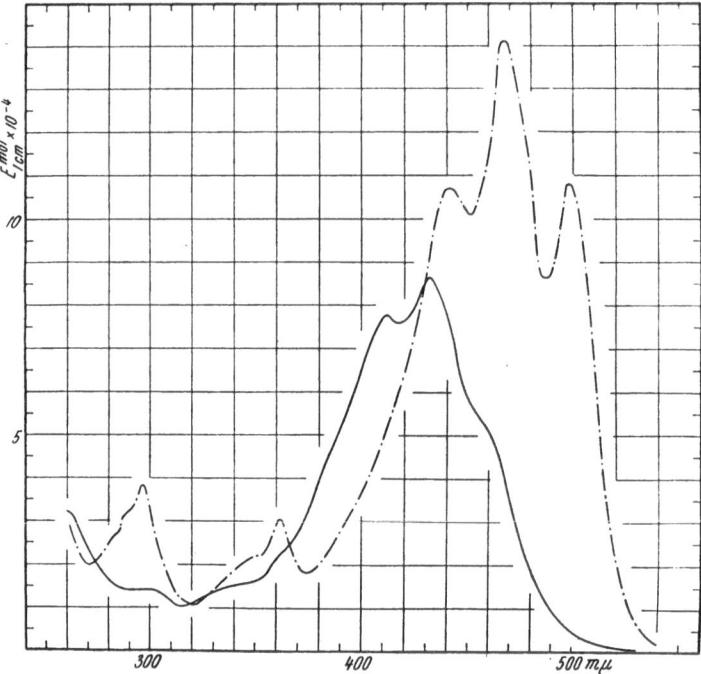

Fig. 20. Molecular extinction curves of a poly*cis*-lycopene (obtained by melting prolycopene crystals or by insolation of solutions), in hexane: ————, fresh solution; and — · — · —, mixture of stereoisomers after catalysis by iodine (*515*). [From: J. Amer. Chem. Soc. **65**, 1940 (1943).]

all-*trans* form as the primary, main rearrangement product. Furthermore, when prolycopene was kept just above its melting point for a few minutes, the subsequent chromatogram was also composed of a dozen colored zones.

Later, in collaboration with MAGOON (*301*), some other stepwise isomerization experiments were conducted by exposing prolycopene to sunshine or artificial light, in the absence of catalysts. These conditions allowed the almost quantitative recovery of the pigment. While one half of the prolycopene remained unchanged and only little all-*trans*-lycopene appeared in the chromatogram, the latter contained relatively large amounts of thirteen "intermediate" *cis* lycopenes, seven of them crystallizable (Nos. 6, 8, 9, 10, 14, 18, and 21 in Table 15, p. 73). One isomer was identical with a natural product, the "poly*cis*-lycopene III" ex *Pyracantha* (cf. Table 13, p. 66). The respective spectra of the "intermediates" showed wide variety with reference to both *cis*-peak intensity and fine structure of the main band. On the column the pigment zones formed distinct groups, probably corresponding to mono-, di- and tri*cis* isomers. This whole picture can be explained exclusively by the poly*cis* nature of prolycopene. Evidently, only in this case can such intermediates appear "en route" to the all-*trans* configuration that are absent from the usual *cis*-*trans*-isomeric equilibria.

This interpretation is also in accordance with pertinent infrared readings (*Fig. 19*, p. 69) [LUNDE et al. (*293*)]. The prolycopene spectrum includes both methylated and unmethylated *cis* double bonds with high intensities (7.25 μ and 13.15 μ; cf. p. 79). In the course of the stepwise isomerization, prolycopene → all-*trans*-lycopene, both bands gradually decrease and finally disappear when the all-*trans* stage is reached (*301*).

In summary, we propose that the prolycopene molecule has a straight overall shape and contains four or five *cis* double bonds. All available data would be explained by a symmetrical penta*cis* model in which the central double bond and four other, sterically unhindered double bonds were *cis*. A confirmation of this working hypothesis could be gained either by the total synthesis of prolycopene or, possibly, by the steric clarification of some partially rearranged pigments. At present, the occurrence of one or two hindered *cis* double bonds in this poly*cis* molecule cannot be excluded.

During the stepwise isomerization a minor isomer (*Fig. 20*) [wrongly named "all-*cis*-lycopene" earlier by this writer (*515*)] appeared *below* the unchanged prolycopene zone. It contains probably one more *cis* double bond than prolycopene. A very similar pigment was observed on feeding prolycopene to chicks (p. 120).

Further Polycis Lycopenes.

As mentioned on p. 72 the fruits of *Pyracantha angustifolia*, grown in Southern California, are unusually

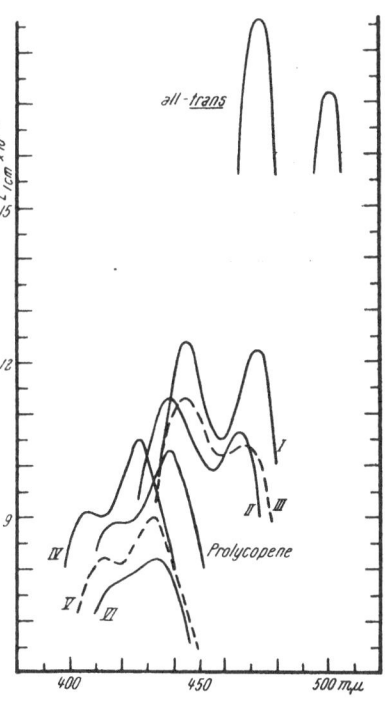

Fig. 21. Molecular extinction curves of all-*trans*-lycopene, prolycopene and poly*cis*-lycopenes I—VI at their main maxima, in hexane (*521*). [From: J. Amer. Chem. Soc. 69, 1930 (1947).]

rich in poly*cis* hydrocarbons [PINCKARD and the writer (*521*)]. Besides prolycopene, six more similar isomers, the "poly*cis*-lycopenes I–VI" were isolated from the ripe berries, and "I"–"III" were crystallized.

Table 14. Spectral Data of Some Poly*cis* Lycopenes ex *Pyracantha* Listed in the Sequence of Decreasing Wavelengths at λ_{max}, and Compared with the All-*trans* and a Mono-*cis* Form (*521*).

Name	Location of λ_{max} in hexane (mμ)	Difference in location of λ_{max} from the all-*trans* form (mμ) (λ_{max} shift)	$E_{1\ cm.}^{mol.} \times 10^{-4}$ at λ_{max}	Planimetrically measured relative extinguished area between 320 mμ and 560 mμ
All-*trans*	472–473	0	18.6	100
Neo A.	465	7.5	12.2	82
Poly*cis* I	444–445	28	12.3	76
Poly*cis* III. . . .	444–445	28	11.3	71
Poly*cis* II.	441	31.5	11.4	70
Prolycopene . . .	438	34.5	10.3	60
Poly*cis* VI	433	39.5	8.1	54
Poly*cis* V	431–432	41	9.0	52
Poly*cis* IV	426	46.5	10.4	62

Although 1 kg. of the fresh berries contained as little as 14 mg. of all-*trans*-lycopene, the following amounts of crystalline *cis* samples were obtained: 46 mg. of prolycopene, 5 mg. of poly*cis*-lycopenes I–III, 13 mg. of pro-γ-carotene (p. 75), and 3 mg. of neo-γ-carotene P (p. 96) — in all, 67 mg. of *cis* forms.

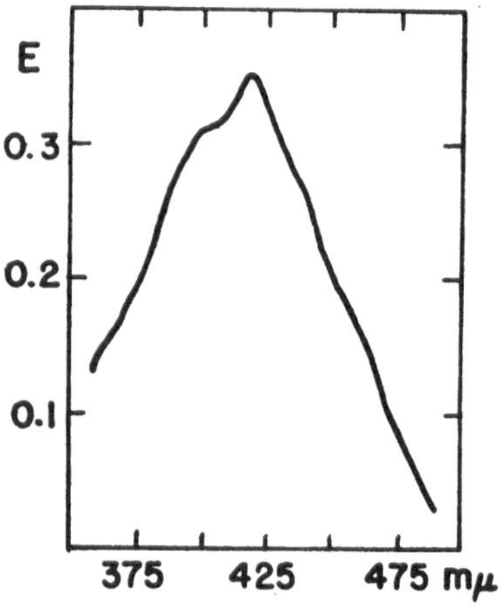

Some spectral data of the poly*cis*-lycopenes I–VI are presented in *Fig. 21* (p. 71) and *Table 14*. In Figs. 37–38 (p. 99) our poly*cis*-lycopenes "III" and "V" are compared with KARRER's synthetic *cis*-lycopenes b, c, and c'.

Table 15 (p. 73) lists also several poly*cis* isomers reported by other research groups. Most of them have not been obtained in crystals, probably because of lack of material. The last pigment mentioned in Table 15 shows a λ_{max} shift of 54.5 mμ and the shortest wavelength maximum (418 mμ) ever recorded in the lycopene set [JOYCE (*218, 219*)] (*Fig. 22*).

Fig. 22. Extinction curve of poly*cis*-lycopene "18" in light petroleum, according to JOYCE (*219*). [From: J. Scient. Food Agric. **10**, 342 (1959).]

Table 15. Stereoisomeric Lycopenes Listed, where Possible, in the Sequence of Decreasing Adsorption Affinities.

Nos. 2–5 are in vitro stereoisomerization products (ex all-*trans*-lycopene); 3–5 are unpublished (the ext. coefficients of No. 2 are higher than reported earlier); 6–8, 12–16, 18–21, 23, 24 are in vitro isomerization products (ex prolycopene); 28–31 are in vivo isomerization products (chick); 33–35 were obtained by total synthesis; and the others were isolated from plants (including No. 18). The relative chromatographic positions of Nos. 11 and 12 are approximate, and those of 28–41 could not be determined by direct comparison. Spectra in hexane or light petroleum. *Italicized* numbers on the left designate crystalline products.

Name of *cis-trans* isomer	λ_{max} (mμ)	$E_{1\ cm.}^{mol.} \times 10^{-4}$	λ_{max} (mμ)	$E_{1\ cm.}^{mol.} \times 10^{-4}$	*cis*-Peak	References
1. All-*trans*..........	472.5	18.6	504.5	17.2	none	(*515*, *536*)
2. Neo A	465	15.1	496	12.9	highest	(*515*, *536*)
3. Neo B	464.5	14.5	495.5	12.3	high	(*515*, *536*)
4. Neo C	461	13.6	492	11.0	high	(*515*, *536*)
5. Neo D	457.5	13.2	488.5	10.6	medium	(*515*, *536*)
6. I A	468.5	15.2	499	12.8	none	(*301*)
7. I B	469	15.1	500	12.6	none	(*301*)
8. I C	464	13.8	495	11.2	medium	(*301*)
9. II A	466.5	13.0	495	9.8	none	(*301*)
10. II B	444.5	12.6	471.5	12.7	none	(*301*)
11. Polycis I.........	444.5	12.3	472	12.4	none	(*521*)
12. "Cryst. isomer"...	445.5	10.2	472	9.7	none	(*515*)
13. II C	439	11.6	464.5	11.8	none	(*301*)
14. II D	439.5	11.0	465	11.8	slight	(*301*)
15. II E	439	10.8	464.5	10.3	small	(*301*)
16. II F	443	11.3	466	10.3	none	(*301*)
17. Polycis II	441.5	11.1	465.5	10.6	none	(*521*)
18. Polycis III	444	11.4	464	10.4	none	(*521*)
19. III B	437	10.5	463	8.8	very slight	(*301*)
20. III C	439.5	9.6	464.5	9.5	small	(*301*)
21. III D	437	10.5	—	—	none	(*301*)
22. Prolycopene	438	10.3	—	—	none	(*516*, *282*)
23. IV B.............	436	9.7	462	9.4	small	(*301*)
24. "All-*cis*" (abandoned term)	432	8.6	—	—	none	(*515*)
25. Polycis IV	426	10.6	—	—	none	(*521*)
26. Polycis V	431.5	8.9	—	—	none	(*521*)
27. Polycis VI	433	8.1	—	—	none	(*521*)
28. Polycis (ex chick).	446	11.0	472	10.4	?	(*89*)
29. Polycis (ex chick).	442	11.0	467	10.0	?	(*89*)
30. Polycis (ex chick).	440	11.0	464	9.9	?	(*89*)
31. Polycis (ex chick).	433	8.9	—	—	?	(*89*)
32. Fraction "1 H" ...	439	—	468	—	?	(*136*)
33. *cis*-Lycopene b....	433	7.4	457	6.9	present	(*128*)
34. *cis*-Lycopene c....	425	10.0	446	9.8	present	(*128*)
35. *cis*-Lycopene c' ...	400–403	7.1	418–420	7.6	?	(*128*)

(continued on next page.)

(Table 15, continued.)

Name of *cis-trans* isomer	λ_{max} (mμ)	$E_{1\ cm}^{mol.}$ × 10⁻⁴	λ_{max} (mμ)	$E_{1\ cm}^{mol.}$ × 10⁻⁴	*cis*-Peak	References
36. Polycis-lycopene "44:66"*	444	—	466	—	none	(218, 219)
37. Polycis-lycopene "37"	437	—	—	—	none	(218, 219)
38. Polycis-lycopene "34"	434–435	—	—	—	none	(218, 219)
39. Polycis-lycopene "32:54"	432	—	454	—	none	(218, 219)
40. Polycis-lycopene "26"	426	—	440–445	—	none	(218, 219)
41. Polycis-lycopene "29"	429	—	—	—	none	(218, 219)
42. Polycis-lycopene "18"	418	—	—	—	none	(218, 219)

* "44–46" means that the maxima are located at 444 mμ and 466 mμ in petroleum ether solution.

Proneurosporene (Protetrahydrolycopene, Neoneurosporene P).

The structure of the all-*trans* compound appears on p. 102.

Proneurosporene, a poly*cis* form, has been observed repeatedly in the fruit of *Lycopersicon* selections and described as "unidentified carotene I" [PORTER and ZSCHEILE (370)], "poly-*cis*-ψ-carotene" [MACKINNEY and JENKINS (297)], and "protetrahydrolycopene" [PORTER and LINCOLN (369); TROMBLY and PORTER (439)]. More recently, MAGOON and the writer (300) called it "neo-neurosporene P" (P for *Pyracantha*). The identity of all these preparations is very likely.

Some tomato selections contained as much as 16 mg. of proneurosporene per kg., besides 26 mg. of prolycopene (369); only 3–4 mg./kg. was present, however, in berries of *Pyracantha angustifolia* (300).

Proneurosporene is absent from stereoisomeric equilibrium mixtures. It does not crystallize but yields, when treated with iodine, the crystalline all-*trans* form (300) which is identical with HAXO's preparation ex *Neurospora crassa* (154).

Spectroscopically, proneurosporene represents an interesting type. Although its fundamental band shows considerable fine structure, with two peaks and two shoulders (Fig. 39, p. 100), the drastic increase of the extinguished area caused by iodine may well be compared with that observed for prolycopene. During this process the 461 mμ maximum doubles its intensity—a feature characteristic of several poly*cis* carotenoids. The absence of a *cis*-peak excludes a bent molecular shape, and the fine structure in the main band excludes the presence of hindered *cis* double bonds. These data are in accordance with the infrared spectrum that does not contain the 13.15 μ band (unmethylated *cis* double bond), while the 7.25 μ band is present with marked intensity (methylated *cis* double bond). The central double bond is very probably *trans*, because of the absence of a band at 12.84 μ which appears in the case of central-mono*cis*-β-carotene but is missing in all-*trans* curves (*293*) (cf. p. 79). These considerations seem to exclude *cis* configurations around four or five of the conjugated double bonds. Hence, we tentatively propose that proneurosporene represents a poly(unhindered)-*cis* type which contains a *trans* CH—CH=CH—CH group in the center [MAGOON and the writer (*300*)].

Upon refluxing, a small fraction of proneurosporene (λ_{max} at 432 mμ) is converted into "neo R" whose maximum extinction (425 mμ) indicates at least one more *cis* double bond than present in the starting material (Fig. 39, p. 100).

Pro-γ-carotene.

This pigment was detected in some palm fruits (Table 13, p. 66) from which only 0.3 mg. of pure crystals were isolated per kg. of fresh material. A practical source of pro-γ-carotene is *Pyracantha angustifolia* whose ripe berries yielded 9 mg. of crystals per kg. [SCHROEDER and the writer (*530*, *531*)].

Pro-γ-carotene forms brick-red glittering plates. Under the microscope the dull brownish-yellow crystals show orange colored crossings. To the naked eye the shade of dilute solutions is similar to that of β-carotene. With reference to thermostability, degraded spectrum, and the dramatic spectral shift upon iodine catalysis, pro-γ-carotene offers a perfect parallel to prolycopene. It yields crystalline γ-carotene when treated with iodine.

The melting point of pro-γ-carotene (135°) cannot be compared directly with that of the all-*trans* form since, for some reason, analytically pure all-*trans*-γ-carotene samples as obtained from various sources show sharp melting points anywhere between 131° and 178° (*531*, *149*, *529*). Furthermore, it has been observed repeatedly that at a certain stage of chromatographic development all-*trans*-γ-carotene formed twin-zones which eventually were washed together.

The spectrum of pro-γ-carotene shows in the visible region two almost equal maxima which are divided by a shallow minimum (Fig. 34, p. 96); this rare feature differentiates pro-γ-carotene from prolycopene. Its configurational meaning is unclear.

The relationship between pro-γ-carotene and neo-γ-carotene P will be mentioned on p. 96.

VIII. Some General Remarks on Configurational Assignments.

As we have seen, the spectral curve of a carotenoid is a function of the molecular shape and hence of the spatial configuration. Were the rules that govern this relationship fully known, it would be easy to assign *cis* configurations. Since this is not the case, some of the following considerations and assignments have tentative character and require confirmation by stereospecific syntheses.

1. Stereoisomeric Types.

As *Table 16* shows, some simple spectroscopic observations can be made in the *cis*-peak and visible regions, before and after catalysis by iodine, that furnish rapid information on the type to which a given *cis* compound belongs. A priori, the following types could appear in stereo-isomeric equilibrium mixtures:

Sterically unhindered forms: central-mono*cis*,
next-to-central-mono*cis*,
peripheral mono*cis*,
di*cis*,
poly*cis*.

Sterically hindered forms: mono-, di-, and poly*cis* (in part also containing unhindered *cis* double bonds).

Table 16. Configurational Types of *normal* Carotenoids as Indicated by Spectroscopic Changes upon Treatment with Iodine.

Configuration	Shape of the curve in the *cis*-peak region	Change of the extinction upon iodine catalysis		Typical example
		in the *cis*-peak region	in the visible region	
All-*trans*	flat	increase	decrease	Fig. 4, p. 27
Central-mono*cis*	high peak	decrease	increase	Fig. 25, p. 81
Peripheral mono*cis*	moderate peak	slight increase	increase	Fig. 24, p. 80
Poly*cis*	flat	increase	strong increase	Fig. 18, p. 68

As was pointed out earlier, the available *trans* → *cis* stereoisomerization methods are highly selective and afford only a few types: (a) With rare exceptions, no sterically hindered isomers appear at equilibrium, and hence no degraded spectra are observed. (b) The isomerized portion of the pigment consists mainly of mono- and di*cis* forms. (c) The incidence of those main *cis* forms that show high *cis*-peaks is remarkably frequent, in spite of the theoretical postulate that only a small fraction of the many possible spatial forms can have *cis*-peaks of high intensity (*515*, cf. p. 41).

In the absence of a theory that would predict the absolute value of the maximum possible *cis*-peak intensity in a given stereoisomeric set, we had assumed earlier that those neo forms that showed the highest *cis*-peaks in the set were central-mono*cis* compounds. However, most of these assignments had to be revised and substituted by next-to-central-*cis* configurations, after the true central-mono*cis* compounds had been synthesized (p. 58). The *cis*-peak intensities of the latter were 10–15% higher than those of the neo forms mentioned.

For rapid characterization of stereoisomers INHOFFEN (*187*) proposed to use the quotient $Q =$ extinction at λ_{max}/extinction at *cis*-peak. This term is also applicable to trace amounts of pigments whose *cis*-peaks cannot be determined quantitatively. The Q value is lowest for central-mono*cis* configurations. Examples: Central-mono*cis*-lycopene, 1.5; central-mono*cis* form of α-carotene, β-carotene, cryptoxanthin, zeaxanthin, or isozeaxanthin, $\sim$ 1.7; next-to-central-mono*cis*-lycopene (neo A), 1.8; and peripheral mono*cis*-β-carotene (neo U), 10.8.

It can now be claimed that in the great majority of our *trans* $\rightleftarrows$ *cis* equilibria central-mono*cis* forms either did not appear at all or perhaps only in trace amounts. This finding might seem to contradict the statement that the activation energy for the *trans* → *cis* rotation about a centrally located double bond is less than for any other double bond in the system, and that this effect may be still further enhanced by the presence, in the middle section only, of the unbranched grouping $=CH—CH=CH—CH=$ (*515*).

This situation was in part clarified by two unexpected properties of the synthetic central-mono*cis* isomers: (a) When compared to other *cis* forms, these compounds showed surprisingly low adsorption affinities and hence trace amounts might have been retained at unexpected locations and escaped detection on the column. (b) It was found [cf. e. g., INHOFFEN et al. (*187*)] that central-mono*cis* configuration involved photosensitivity, even in scattered daylight. Future experiments, to be conducted in red light, might still afford some representatives of this theoretically important type as products of direct isomerization. Central-mono*cis*-canthaxanthin has been obtained even in daylight.

Configurational assignments become difficult when one or more sterically hindered *cis* double bonds are present. As indicated on p. 78, one cannot differentiate between poly*cis* and hindered mono*cis* types by comparing the fundamental bands before and after a treatment with iodine; upon catalysis both types show a spectacular increase of the extinguished area. A differentiation may be feasible, however, in the *cis*-peak region because the presence or absence of a *cis*-peak is not determined by the hindered or unhindered nature of *cis* double bonds but by the shape of the chromophore. Whereas the overall straight form of a poly*cis* molecule does exclude the presence of a *cis*-peak, a more or less marked peak may appear in the case of a bent molecule that contains hindered *cis* double bond(s) (cf. Figs. 37, 38, p. 99).

That a *cis*-peak and a degraded main band may occur in the same spectrum is also illustrated by the curve of a *cis-retro*-dehydrocarotene that was prepared by melting all-*trans* crystals [WALLCAVE et al. (537)].

The distinction of poly*cis* and hindered mono*cis* carotenoids by the method of stepwise isomerization was discussed on p. 7ɔ. We cannot differentiate, however, between poly*cis* compounds that contain, respectively, only unhindered, only hindered, or both hindered and unhindered *cis* double bonds, except perhaps by IR spectra.

2. Number and Location of *cis* Double Bonds.

Most attempts to rearrange all-*trans* carotenoids have yielded only one or two main *cis* forms whose interpretation as mono- or di*cis* isomers was reasonable considering the high probability of their formation. Furthermore, numerous spectroscopic readings have revealed that in a number of instances λ_{max} of the main neo form(s) is located at a 5 mμ ($\pm$ 1 mμ) shorter wavelength than that of the corresponding all-*trans* compound (in hexane, petroleum ether, cyclohexane or benzene solution) (492). It is safe to assume that this consistently observed minimum value of the λ_{max} shift indicates a mono*cis* configuration. Other main isomers which displayed a shift of $\sim$ 10 mμ were classified as di*cis* forms, etc. As the following remarks will show, the reliability of such assignments decreases rapidly when the λ_{max} shift exceeds 10 mμ.

If in a long, conjugated chain all sterically unhindered double bonds were energetically equivalent, then any single *trans* → *cis* rotation would have the same, additive spectral effect and the number of the *cis* double bonds present would simply follow from the position of λ_{max}. Evidently, this is not the case. As is well known, each double bond in the polyene molecule loses an individual fraction of its double bond character to the adjoining single bonds, and the amount thus lost increases from the ends toward the center of the conjugated system. Consequently, among the various ionic structures mentioned on p. 38, there are more which give partial single bond character to the double bonds at or near the center than near the ends [COULSON (72), cf. (515)]. The individual (unhindered) *cis* double bonds produce different λ_{max} shifts and these are not necessarily additive.

The correctness of the above considerations was confirmed by recent total syntheses of the central-mono*cis* forms of α-carotene, β-carotene, 3,4-dehydro-β-carotene, 3,4,3',4'-bisdehydro-β-carotene, cryptoxanthin, zeaxanthin, isozeaxanthin, etc. (p. 58): In all these instances the λ_{max} shift amounted to only 3 mμ, and for lycopene to 2 mμ (187)—a clear demonstration of the decrease of the λ_{max} shift value as we proceed from an end of the conjugated system towards the center.

Another spectral feature that is dependent on the number of *cis* double bonds is the degree of fine structure in the fundamental band. As mentioned before, the fine structure decreases upon any *trans* → *cis* rotation (p. 27). The contributions of individual *cis* double bonds to this flattening effect may well be additive but, unfortunately, no quantitative use can be made of this relationship at the present time.

3. Configuration and Infrared Spectrum.

The application of this powerful tool in the field of *cis-trans*-isomeric carotenoids is still in an early stage. Most of the pertinent reports are concerned with shorter conjugated systems, e. g. *cis-cis*, *cis-trans*, and *trans-trans* dienes [RASMUSSEN et al. (*383, 385, 386*); SHEPPARD et al. (*405, 404*); JACKSON et al. (*211*); PASCHKE et al. (*345*); CELMER et al. (*57*); SZASZ et al. (*431*)].

A modern survey of this field has been presented by BRAUDE and WAIGHT (*41*); cf. also ALLAN et al. (*7*), COLE (*67*), and writer (*498*).

LUNDE and the writer (*291*) have found in the diphenylpolyene series (p. 160) that the influence of the configuration on the IR curve is manifest in the following regions: 7.0–7.1 μ (in-plane vibration of CH being part of a *cis* C=C double bond); 12.84–12.95 μ (the analogous out-of-plane vibration); and 10.0–10.6 μ (out-of-plane vibration of CH in the corresponding *trans* grouping). In the class of the carotenoid pigments the stereochemically sensitive regions are, ~ 7.25 μ, ~ 13 μ, and 10.0–10.6 μ (example, *Fig. 23*).

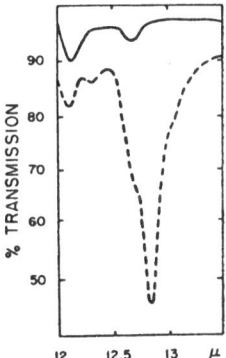

Fig. 23. Infrared spectra curves in cyclohexane (1 mm. NaCl cell): ——, 0.5% solution of all-*trans*-β-carotene; and – – –, 1.0% solution of synthetic central-mono*cis*-β-carotene (*498*). [From: Experientia 10, 1 (1954).]

Because of the absence of aromatic vibrations, the IR curves of carotenes are simpler than those of diphenylpolyenes (*291*). On the other hand, the isoprenic distribution of methyl groups in the carotenoid molecule creates two types of olefinic bonds, conveniently termed "methylated" —(CH$_3$)C=CH— and "unmethylated" —CH=CH— double bonds. Thus, in the β-carotene molecule four aliphatic double bonds are methylated and five unmethylated. One of the latter occupies an exceptional position at the center, i. e. in the sole section of the molecule where neither of four consecutive carbon atoms carries a side-chain: —CH=CH—CH=CH—.

Cis configurations of methylated and of unmethylated double bonds cause different IR effects. Some pertinent data will be given in the following Chapter for the β-carotene set and a few other sets.

IX. Configurational Assignments in Certain Stereoisomeric Sets.

In this Chapter an attempt will be made to summarize our knowledge concerning *cis* configurations proposed on the basis of stereoisomerization experiments and to coordinate them with the information gained by total synthesis. The reader will note that the available data are both heterogeneous and fragmentary.

1. Stereoisomeric Sets with Two Hydroaromatic Terminal Groups.

β-Carotene Set.

Structure, p. 5.

Cis isomers obtained by rearrangement, GILLAM and EL RIDI (*129, 130*); POLGÁR and the writer (*365, 524*).

Cis isomers prepared by total synthesis, INHOFFEN et al. (*185, 187*); ISLER et al. (*203, 195*); EUGSTER, GARBERS and KARRER (*107*).

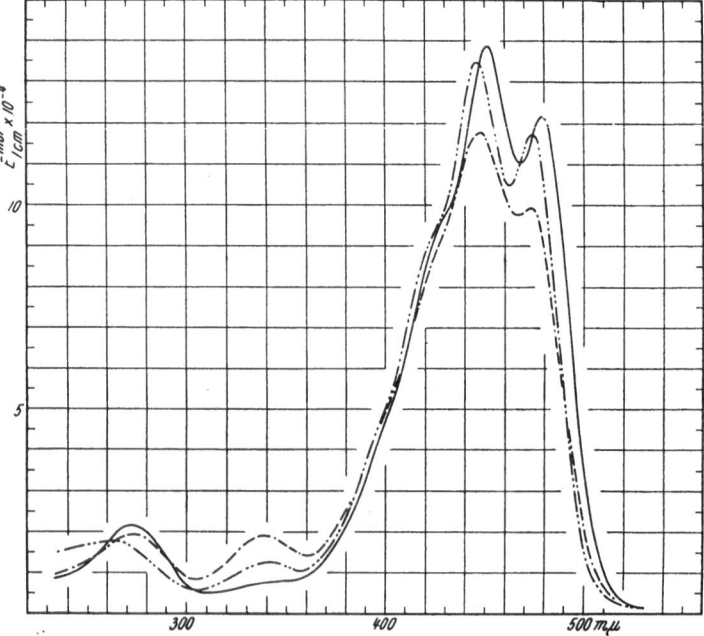

Fig. 24. Molecular extinction curves of neo-β-carotene U, compared with that of all-*trans*-β-carotene, in hexane: ————, all-*trans* form; —··—, fresh solution of neo-β-carotene U; and —·—, mixture of stereoisomers obtained from neo U with iodine (*524*). [From: J. Amer. Chem. Soc. **65**, 1522 (1943).]

Stereoisomerization of all-*trans*-β-carotene has yielded 12 *cis* forms, among them 2 main isomers (20–25% each), viz. neo U (adsorbed above the all-*trans* pigment) and B (adsorbed below it). Considering the UV spectra *(Fig. 24)* neo U was interpreted as a peripheral 9-mono*cis*-β-carotene (Model II, p. 14; λ_{max} shift, 5.5 mμ, distinct but low *cis*-peak, considerable thermostability), while for neo B (GILLAM's ψ-α-carotene; λ_{max} shift, 10.5 mμ, high *cis*-peak, marked lability; partial rearrangement during crystallization) the 9,15-di*cis* configuration was tentatively pro-

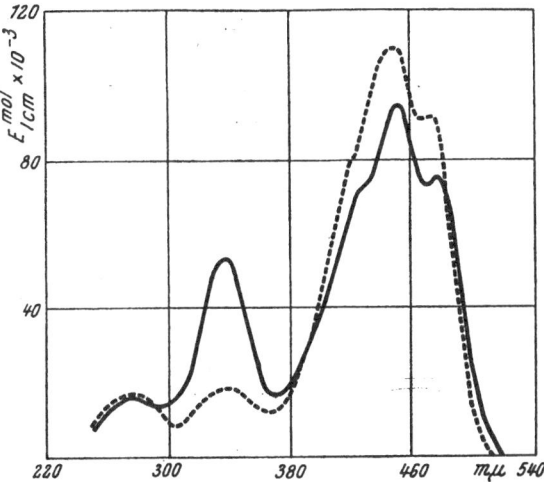

Fig. 25. Molecular extinction curves, in light petroleum: ————, fresh solution of synthetic central-mono*cis*-β-carotene; and — — —, mixture of stereoisomers after catalysis by iodine; according to INHOFFEN et al. (*185*). [From: Liebigs Ann. Chem. 570, 54 (1950).]

posed. Recently, these isomers were compared with INHOFFEN's synthetic central-mono*cis* compound (Model IV, p. 14; λ_{max} shift, 3 mμ) which has the highest *cis*-peak in this set, viz. $\varepsilon = 5.2 \times 10^4$ *(Fig. 25)*. It is crystalline but shows photolability (p. 77).

With reference to the IR spectra *(Fig. 26)* the following points can be made [LUNDE and the writer (*293*)]. Within the 10.0–10.6 μ region certain bands had been assigned by earlier authors (*405*) to out-of-plane vibrations of the two H-atoms in a *trans* C—CH=CH—C grouping. The latter group occurs five times in the all-*trans*-β-carotene, neo U, and neo B molecules but only four times in the central-mono*cis* isomer. Indeed, the 10.35 μ band appears in the latter curve with markedly lower intensity than in the three others. Furthermore, it has been known that conjugation of the mentioned group with a similar *cis* grouping, to form a *trans-cis* diene, causes the splitting of this band into a doublet or triplet (*211, 57, 404, 385, 478*). We did observe such a split when working

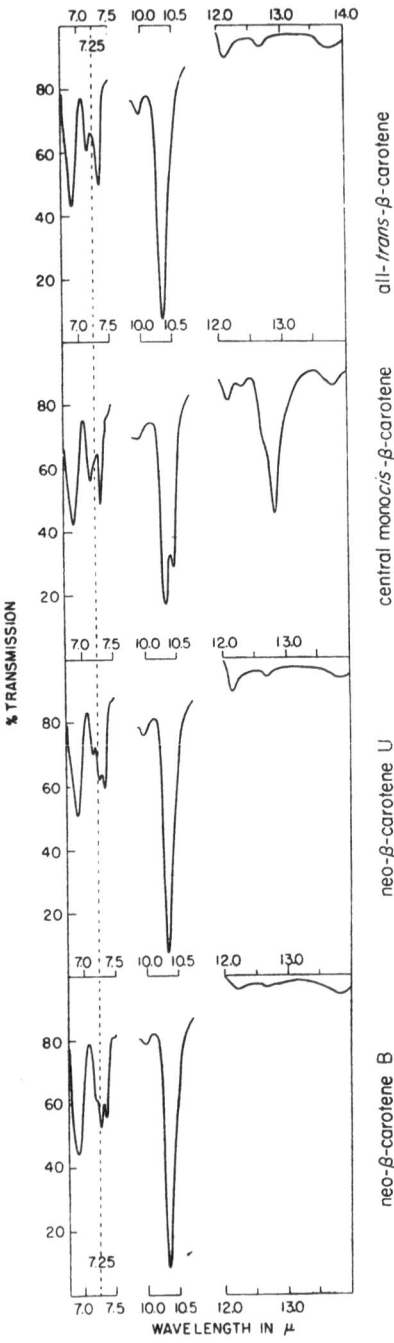

with lower-molecular diphenylpolyenes but predicted that it might
be obscured in more extended conjugated systems (*291*). In the
central-mono*cis* - β - carotene curve,
however, a doublet clearly appears
(at 10.35 μ and 10.47 μ, Fig. 26) as
confirmed by ISLER et al. (*203*).

The neo-β-carotene U and B
curves show a distinct band at
7.25 μ that as a rule is missing in
central-mono*cis*- and all-*trans*-caro-
tenoid spectra. We attribute this
band which is located within the
region assigned earlier (*384*) to
deformation vibrations of methyl
groups, to such vibrations in a
methylated cis double bond.

This "rule" is not an absolute one.
Thus, the all-*trans*-α-carotene curve
does show a very slight peak at
7.25 μ (interpretation on p. 92; cf. also
p. 89).

Central-mono*cis*-β-carotene exhibits a strong band at 12.84 μ that
is absent from the all-*trans*, neo U
and neo B curves. This band has
also been observed in the curves of
the central-mono*cis* forms of methylbixin and dimethylcrocetin (pp. 109,
117) and indicates the presence of
unmethylated cis double bonds. It
is not expected to appear in the
spectra of such carotenoids which
either do not possess *cis* double
bonds at all (all-*trans* forms) or

Fig. 26. Stereochemically important sections taken
from infrared curves of some *cis-trans*-isomeric β-caro-
tenes: 1.0% CCl$_4$ solutions (1 mm. cell) in the
7.0—7.5 μ and 10.0—10.5 μ regions; and 0.5% cyclohexane solutions (1 mm. cell) in the 12.0—14.0 μ
region, except for the central-mono*cis* isomer whose
solubility in this solvent permitted the use of a
1.0% solution (*293*). [From: J Amer. Chem. Soc.
77, 1647 (1955).]

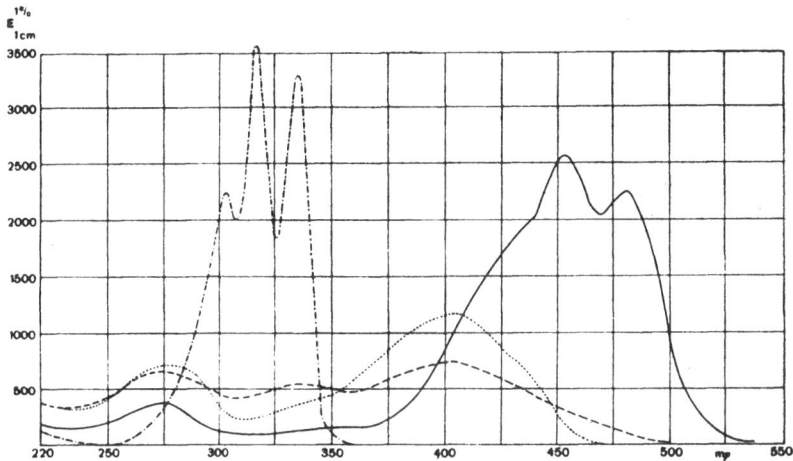

Fig. 27. Specific extinction curves, in light petroleum: — — —, synthetic (hindered) 11,11'-di*cis*-β-carotene; ————, all-*trans*-β-carotene; · · · ·, 11,12,11',12'-bisdehydro-β-carotene; and — · — · —, 3,8-dimethyl-decatriene-(3,5,7)-diyne-(1,9); according to ISLER et al. (*195*). [From: Helv. Chim. Acta 40, 1256 (1957).]

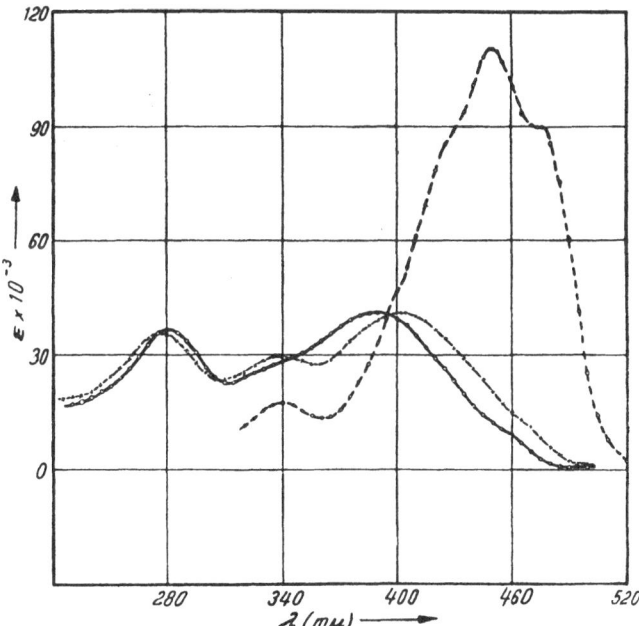

Fig. 28. Molecular extinction curves, in cyclohexane: — × — × —, synthetic (hindered) *cis*-β-carotene B; —o—o—, synthetic (hindered) *cis*-β-carotene C; and the highest curve, same after catalysis by iodine; according to EUGSTER, GARBERS and KARRER (*107*). [From: Helv. Chim. Acta 36, 1378 (1953).]

6*

contain methylated *cis* double bonds only. These findings confirm the 9-mono*cis* configuration of neo-β-carotene U which cannot contain a central *cis* double bond and must possess a methylated *cis* double bond. Hence, *cis* configurations at 7, 11, 15, 7', and 11' are excluded (numbering, p. 5). The 13-*cis* (and 13'-*cis*) configurations are excluded too because they would involve high *cis*-peaks (cf. Model III, p. 14); thus, the 9-mono*cis* configuration is vindicated.

The neo-β-carotene B curve shows the presence of a methylated *cis* double bond (7.25 μ) and the absence of unmethylated such bonds (12.0–14.0 μ). Hence, both *cis* double bonds of neo B must be methylated. Clearly, our earlier assumption (induced by the relatively high *cis*-peak) of the presence of a central-*cis* double bond had to be abandoned. *Cis* configurations about the other four unmethylated double bonds (7, 7', 11, and 11') are also excluded because they represent sterically hindered types which would be incompatible with the spontaneous formation of neo B from all-*trans*-β-carotene and with the non-degraded fundamental band. This leaves only the double bonds 9, 9', 13, and 13' available for *cis* assignments, i. e. the di*cis* configurations 9,9'; 9,13; 13,13'; and 9,13' (Models VIII, V, X, and VII on p. 14). Since the three former configurations would involve slight or no *cis*-peaks, we propose that neo-β-carotene B is 9,13'-di*cis*-β-carotene (Model VII).

The sterically hindered 11,11'-di*cis*-β-carotene (*cis*-β-carotene B) was synthesized by EUGSTER, GARBERS and KARRER (*107*) and by ISLER et al. (*195*). As required by the configuration (**XXXV**) this compound shows a degraded fundamental band (*Fig. 27—28*, p. 83).

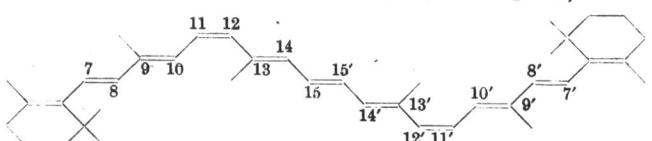

(**XXXV.**) 11,11'-Di*cis*-β-carotene (both *cis* double bonds are sterically hindered).

In the IR region two unexplained bands appear (13.13 μ, 13.50 μ) (*195*) which do not occur in the spectra of the known unhindered *cis* β-carotenes; however, a strong 13.15 μ band is shown by prolycopene.

The synthetic, crystalline "*cis*-β-carotene C" may possess, besides the sterically hindered 11- and 11'-*cis* double bonds, some unhindered double bonds adjacent to the hindered ones (*107*) (*Fig. 28*, p. 83).

Cryptoxanthin (β-Cryptoxanthin) Set.

Structure, 3-hydroxy-β-carotene, p. 6.

Cis isomers obtained by rearrangement, LEMMON, CHATTERJEE and the writer (*512*, *59*); CHOLNOKY and SZABOLCS (*61*).

Cis isomer found in nature, CURL (*77*).

Cis isomers prepared by total synthesis, ISLER et al. (*202*).

Upon iodine catalysis of all-*trans*-cryptoxanthin, the mixture contained two main, crystallizable *cis* isomers, neo U and neo B (yield of each one, 18%; λ_{max} shifts, 5 mμ and 4.5 mμ). These two mono*cis* compounds are easily differentiated by their *cis*-peaks. In the case of neo U (adsorbed above the all-*trans* form) the moderate height of the peak is very similar to that of neo-β- or -α-carotene U, hence the *cis* double bond must be located peripherally. Because of the non-symmetrical structure, neo-cryptoxanthin U may represent either the 9-*cis* or the 9'-*cis* form. For neo B the molar extinction coefficient at *cis*-peak (4.5 $\times$ 10^4) is almost identical with the value found for neozeaxanthin A but lower than that for ISLER's (amorphous) central-mono*cis*-cryptoxanthin (4.8 $\times$ 10^4). We assign to neocryptoxanthin B the next-to-central-mono*cis* configuration (13- or 13'-*cis*).

In the minor isomer, neocryptoxanthin A, probably two double bonds have assumed *cis* configurations (λ_{max} shift, 6.5 mμ).

Zeaxanthin Set.

Structure, 3,3'-dihydroxy-β-carotene (p. 6).
Cis isomers obtained by rearrangement, CHOLNOKY, POLGÁR, TUZSON, LEMMON and the writer (*504, 536, 512*).
Cis isomers prepared by total synthesis, ISLER et al. (*201, 207*).

Both main, crystallizable isomers, neo A and neo B (yields, 10% and 20%), show a λ_{max} shift of 5.5 mμ. On the basis of the *cis*-peak intensities they have been assigned earlier the 15- and 13-mono*cis* configurations, respectively. In the IR region both *cis* curves (but not the all-*trans* curve) show, with very similar intensities, a band at 7.25 μ which is attributed to a methylated *cis* double bond.

The minor differences between the all-*trans* and either of the two *cis* curves in the 12.0–14.0 μ region could not possibly prove the presence of an unmethylated *cis* double bond. This is confirmed by the equal intensities of the *trans* double bond maxima at 10.35 μ in all three curves.

The proposed mono*cis* nature of the neozeaxanthins A and B is in accordance with the IR data; and the observed absence of unmethylated *cis* double bonds excludes *cis* configuration about the central double bond. Furthermore, the *cis*-peak of neo A, although the highest among all isomers obtained by rearrangement of all-*trans*-cryptoxanthin, is much lower than that of ISLER's synthetic central-mono*cis*-zeaxanthin. Hence, we propose that neo A is 13-mono*cis*-zeaxanthin (next-to-central-*cis* form). If so, then the only possible assignment for neo B is that of 9-mono*cis*-zeaxanthin, considering the postulate that its *cis* double bond must be methylated.

No IR data are available for the minor, crystalline di*cis* isomer neo C (λ_{max} shift, 8.5 μ); its considerable *cis*-peak indicates a bent molecular form.

Isozeaxanthin Set.

Structure, 4,4'-dihydroxy-β-carotene (p. 6).

Cis isomers prepared by total synthesis, ISLER et al. (*205*).

The only reported stereoisomer belonging to this set is the synthetic, crystalline (meso or racemic) central-mono*cis*-isozeaxanthin that exists in two unclarified modifications, a and b. The λ_{max} shift amounts to 2–3 mμ.

Echinenone Set.

Structure, 4-keto-β-carotene.

Cis isomer prepared by total synthesis, AKHTAR and WEEDON (*5*).

The central-mono*cis* form (λ_{max} at 454 mμ) shows a λ_{max} shift of 2 mμ (in petroleum ether).

Canthaxanthin Set.

Structure, 4,4'-diketo-β-carotene (p. 6).

Cis isomers obtained by rearrangement, GANSSER and the writer (*123*).

Cis isomers prepared by total synthesis, ISLER et al. (*205, 539*).

This fungus pigment yielded on rearrangement the neo compounds A–F of which A, B, and C crystallized. They show decreased adsorption affinities. Although central-mono*cis* forms do not occur as a rule in stereochemical equilibria, the minor isomer, neo C is identical with ISLER's totally-synthetic central-mono*cis*-canthaxanthin. Whereas the λ_{max} shift of central-mono*cis*-β-carotene is only 3 mμ, the corresponding value in the canthaxanthin set is 4–5 mμ (in hexane).

The two main neo forms A and B (yields, 18% and 10%) show the respective λ_{max} shifts, 7 mμ and 9 mμ in hexane, and 6 mμ and 10 mμ in benzene *(Figs. 29–30)*. Neo A represents a mono*cis* form and, considering the flatness of its curve in the *cis*-peak region, we assign to it the 9-mono*cis* configuration. The next-to-central-mono*cis* configuration (13-*cis*) is tentatively proposed for neo B in spite of its considerable λ_{max} shift, because its *cis*-peak is but slightly inferior (by 13% in hexane, by 8% in benzene) to that of the central-mono*cis* compound.

As expected, the IR spectrum shows a band at 12.84 μ (unmethylated *cis* double bond) only in the central-mono*cis* curve.

The intense bands at 13.15–13.20 μ in the neo A, B, and C curves must have some stereochemical significance that cannot be defined at the present time; no such band is present in the all-*trans* curve.

In the neocanthaxanthins D, E, and F more than one (probably two) *cis* double bonds are present (λ_{max} shifts in hexane 14–18 mμ, and in benzene 16–18 mμ). Since the curves are flat in the *cis*-peak region, neither of these isomers can possess a bent molecular shape.

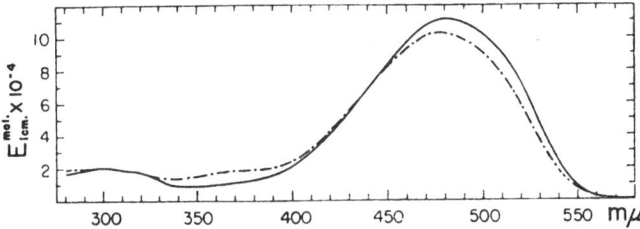

Fig. 29. Molecular extinction curves of canthaxanthin, in benzene: ————, all-*trans* compound; and — · — · —, mixture of stereoisomers after catalysis by iodine (*123*). [From: Helv. Chim. Acta 40, 1757 (1957).]

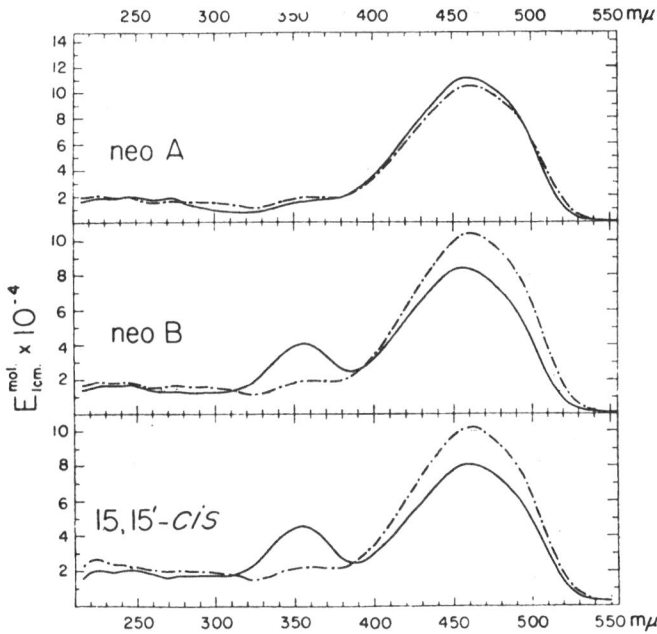

Fig. 30. Molecular extinction curves of some neocanthaxanthins in benzene: ————, neo A, B, and C (the latter is identical with the synthetic 15,15′-*cis* form); and — · — · —, mixture of stereoisomers after catalysis by iodine (*123*). [From: Helv. Chim. Acta 40, 1757 (1957).]

β-*Carotene Monoepoxide Set.*

Structure, p. 6.
Cis isomers obtained by rearrangement, TSUKIDA and the writer (*440*).

Among the nine *cis* forms observed on iodine catalysis the neo-epoxides W (yield, 14%), T (15%) and B (7%) are preponderant. W and T show moderate *cis*-peaks; that of B is higher but not the highest in the set. In all three instances the λ_{max} shift amounts to 6–8 mμ. It is difficult to assign configurations in this unsymmetrical set. We believe that neo T, the most abundant isomer, has a peripheral mono*cis* configuration (9- or 9'-*cis*) because of the flatness of its curve in the *cis*-peak region. The much reduced fine structure of the neo D and E curves in the fundamental band is remarkable. The two curves are similar in this respect, although the λ_{max} shifts are very different (5.5 mμ and 13 mμ).

Some of the *cis* epoxides show in the 12.8–13.2 μ region strong bands whose evident stereochemical significance is unclear. Concerning the occurrence of the 7.25 μ band in the all-*trans* curve a tentative explanation is given on p. 92.

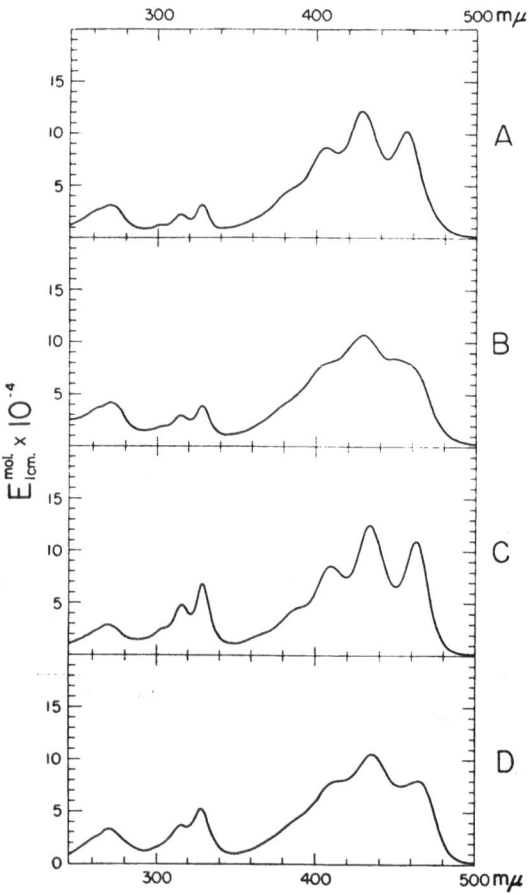

Fig. 31. Molecular extinction curves of the neo-β-carotene diepoxides A. B, C, and D, in hexane (*440*).
[From: Arch. Biochem. Biophys. **74**, 408 (1958).]

β-Carotene Diepoxide Set.

Structure, p. 6.

Cis isomers obtained by rearrangement, Tsukida and the writer (*440*).

The all-*trans* form yielded seven *cis* isomers; the two main ones, neo U (25%) and neo C (10%), are mono*cis* forms (λ_{max} shifts, 4 mμ and 5.5 mμ). For neo U (adsorbed above the all-*trans* form; low *cis*-peak) the 9-*cis* configuration is proposed, while the extraordinarily high *cis*-peak of neo C ($\varepsilon = 6.8 \times 10^4$) would be indicative of the central-*cis* configuration (*Fig. 31*, p. 88). Nevertheless, because of the absence of unmethylated *cis* double bonds, the next-to-central-mono*cis* (13-*cis*) configuration is tentatively assigned to this isomer.

No configuration can yet be proposed for the mono*cis* from neo D (λ_{max} shift, 3.5 mμ; high *cis*-peak). The minor di*cis* form, neo V (λ_{max} shift, 8.5 mμ) must have a straight molecular shape because of the flatness of its curve in the *cis*-peak region.

Unexpectedly, the 7.25 μ band did appear in the IR spectra of the two all-*trans*-β-carotene epoxides. This band is believed to be caused by deformation vibrations of CH$_3$ attached to the epoxide group. Hence, the situation is "*cis*-like" as at the α-ionone end of the α-carotene molecule where a methyl is connected with an isolated ring double bond (cf. p. 92). This explanation seems to be vindicated by the markedly higher intensity of the 7.25 μ band in the β-carotene diepoxide than in the monoepoxide curve.

The strong bands located at 12.75 μ and 13.05 μ in the neo U and V curves, as well as the 13.05 μ band in the neo B curve, are missing in the spectra of the neo forms A, C, and D. Their evident stereochemical significance is unclear.

Luteochrome Set.

Structure, β-carotene-epoxide-furanoid oxide; composition $C_{40}H_{56}O_2$.

Cis isomers obtained by rearrangement, Suzuki and Tsukida (*430*).

By the usual treatments of the all-*trans* compound four *cis* forms were obtained, none of which crystallized. Two of them were adsorbed above and two below the all-*trans* zone. The spectral data point at mono*cis* configurations. The neo U and all-*trans* spectra are practically identical. During the steric rearrangement some structural isomerization to the corresponding difuranoid oxide, aurochrome has also occurred.

3,4-Dehydro-β-carotene and 3,4,3',4'-Bisdehydro-β-carotene Sets.

Structures, p. 6.

Cis isomers obtained by rearrangement, Karmakar and the writer (*221*).

Cis isomers prepared by total synthesis, Isler et al. (*204*); Inhoffen and Raspé (*191*).

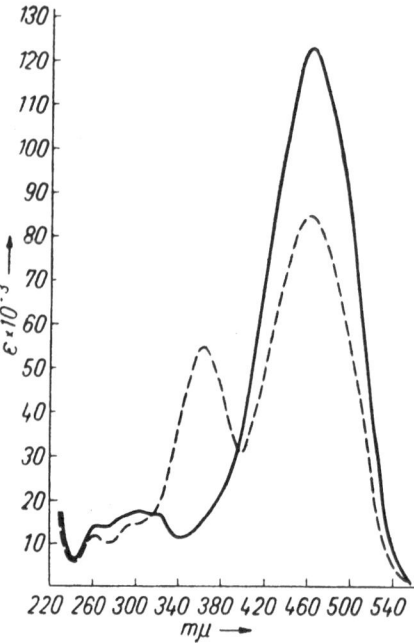

Fig. 32. Molecular extinction curves of 3,4,3',4'-bisdehydro-β-carotene, in light petroleum: ———, all-*trans* form; and — — —, central-mono*cis* form; according to INHOFFEN and RASPÉ (*191*). [From: Liebigs Ann. Chem. **594**, 165 (1955).]

All-*trans*-3,4-dehydro-β-carotene afforded on iodine catalysis the two main isomers neo U (yield 22%; adsorbed above the all-*trans* form; λ_{max} shift, 7 mμ; slight *cis*-peak) and neo A (27%; adsorbed below all-*trans*; λ_{max} shift, 11 mμ; considerable *cis*-peak). They were interpreted, respectively, as a peripheral mono*cis* and a di*cis* isomer, the latter having a bent molecular shape.

The totally-synthetic central-mono*cis*-3,4-dehydro- and 3,4,3',4'-bisdehydro-β-carotenes (λ_{max} shifts, 4 mμ) show as expected very high *cis*-peaks (cf. *Fig. 32*).

16,16'-Homo-β-carotene Set.

Structure, p. 41; composition $C_{42}H_{58}$.

Cis isomers prepared by total synthesis, INHOFFEN et al. (*183*).

The configurations of the middle-mono*cis* (15-*cis*) and middle-di*cis* (15,15'-di*cis*) isomers (λ_{max} shifts, 6 mμ and 9 mμ) are represented on p. 41, where the respective *cis*-peaks are also discussed.

13,13'-Bis-desmethyl-β-carotene Set.

Structure, see below; composition $C_{38}H_{52}$.

Cis isomers prepared by total synthesis, INHOFFEN, BOHLMANN and RUMMERT (*186*).

The main, crystalline *cis* form **a** shows a high *cis*-peak and a λ_{max} shift of only 2 mμ; it represents the central-mono*cis* compound. The curve of the minor isomer *b* is almost flat in the *cis*-peak region.

$$\left[\begin{array}{l} \underset{\displaystyle CH_2 \diagdown \underset{\displaystyle CH_2}{\overset{\displaystyle |}{C}}-CH_3}{\underset{\displaystyle |}{CH_2}} \diagup \overset{\displaystyle CH_3 \diagdown \underset{\displaystyle C}{\diagup} CH_3}{C} \\ CH_2 \quad C-CH=CH-C=CH-CH=CH-\overset{13}{CH}=CH-CH= \end{array} \right]_2$$

(XXXVI.) 13,13'-Bis-desmethyl-β-carotene.

retro-Dehydrocarotene Set.

Structure, p. 6.

Cis isomers obtained by rearrangement, WALLCAVE and the writer (*537, 475*).

Whereas *normal* carotenes, when submitted to iodine catalysis, retain about one half of their molecules in the all-*trans* configuration, the corresponding value for this *retro* compound is as low as $^1/_6$. Among the observed eight *cis* forms which all appeared below the all-*trans* zone, two main, crystallizable isomers were preponderant, viz. neo A (yield, 28%; λ_{max} shift, 4 mμ) and neo D (12%; 7 mμ); they represent a mono- and a di*cis* form. Their molecular shape is obscure since no definite *cis*-peaks are observable in this set (cf. p. 34).

When fused, all-*trans-retro*-dehydrocarotene was converted to the extent of 1%ʼ into neo J (λ_{max} shift, 22 mμ). This is an interesting, hindered *cis* isomer with a degraded spectrum; the extinction at λ_{max} amounted to only 63% of that of the all-*trans* compound.

α-Carotene Set.

Structure, p. 5.

Cis isomers obtained by rearrangement, GILLAM, EL RIDI and KON (*131*); POLGÁR and the writer (*525*).

Cis isomer prepared by total synthesis, INHOFFEN, SCHWIETER and RASPÉ (*192*); see also ISLER et al. (*394a*).

This structurally and sterically unsymmetrical carotenoid contains ten conjugated double bonds and hence two non-equivalent "middle" double bonds, one of which is located at the same place in the chain as is the central double bond of β-carotene, while the other "middle" double bond is equivalent to a next-to-central double bond in β-carotene. Iodine catalysis afforded ten isomers, among them the three main forms, neo U (yield, 9%; adsorbed above the all-*trans* zone; crystalline; λ_{max} shift, 5.5 mμ), neo W (8%; crystalline; 6.5 mμ), and neo B (13%; 10.5 mμ); the latter represents a di*cis* compound. Like neo-β-carotene U, neo-α U is considerably thermostable. It shows a moderate *cis*-peak and is to be interpreted as a peripheral mono*cis* isomer, viz. 9-*cis*-α-carotene (the 9'-*cis* configuration cannot be excluded). Our assignment is confirmed by IR readings: a band appears at 7.25 μ (methylated *cis* double bond), while the 12.0–14.0 μ band is missing. This eliminates *cis* configurations about the unmethylated double bond in position 15 (those in 7, 11, 11', and 7' are hindered).

The *cis*-peak of neo-α-carotene B is much higher than that of neo U. Neo B was at first tentatively assigned the 13,9'- or 15,9'-di*cis* configuration. The IR data exclude, however, the presence of un-methylated *cis* double bonds and hence the 15,9'-di*cis* configuration.

We believe that neo-α-carotene B is the 13,9'- (or possibly the 9,13'-) di*cis* isomer (similar to Model VII on p. 14).

The neo-α- and -β-carotenes B on one hand, and the two neo U forms on the other, seem to have analogous configurations; within the two corresponding pairs they show practically equal *cis*-peak intensities.

A special feature of all-*trans*-α-carotene is a slight but distinct maximum at 7.25 μ which can be explained as follows [LUNDE et al. (*293*)]: In general, this band is brought about by deformation vibrations of a methyl group attached to a *cis* double bonded carbon atom in an open chain. In α-carotene a methyl group is attached to an isolated double bond that is located in a cyclohexene ring and has "*cis*-like" character. Since the influence of such a double bond on the vibrations of the methyl group must be similar to that of an aliphatic *cis* double bond, a 7.25 μ band was expected and found in the all-*trans*-α-carotene curve [cf. FARRAR et al. (*115*)].

Somewhat similar is the situation in case of the β-carotene epoxides (pp. 87—88). It should also be mentioned that the inflection at 7.2 μ is much more distinct in the α-ionone than in the β-ionone curve (*400*).

There exist also other "*cis*-like" situations (cf. p. 13).

In the α-carotene set the intensities at 7.25 μ increase in the sequence, all-*trans* < neo U < neo B, indicating the respective presence of 0, 1, and 2 *aliphatic* methylated *cis* double bonds.

The crystalline 15,15'-mono*cis* form of (rac.) α-carotene ("central" with reference to the carbon skeleton but not to the conjugated system) was synthesized by INHOFFEN. Possibly, our minor isomer neo-α-carotene C (yield, 1%) is identical with this compound, except for the rotatory power (*cis*-peaks, 4.4 and 4.6 $\times$ 10^4).

Neo-α-carotene X, a di- (or tri-) *cis* isomer (λ_{max} shift, 13.5 mμ) is formed preponderantly upon mild heating of neo U solutions. It is reasonable to assume that the location of one of its *cis* double bonds is identical with that of neo U (9- or 9'-position).

No assignment can yet be proposed for neo-α-carotene W.

α-Cryptoxanthin (Physoxanthin) Set.

Structure, 3'-hydroxy-α-carotene (p. 7).
Cis isomers obtained by rearrangement, CHOLNOKY et al. (*62*); cf. BODEA et al. (*37, 38*).

The three observed *cis* forms, neo U, neo A, and neo B, showed the respective λ_{max} shifts, 4 mμ, 7 mμ, and 6 mμ. Neo-α-cryptoxanthin U adsorbs above the all-*trans* zone.

Lutein Set.

Structure, 3,3'-dihydroxy-α-carotene (p. 7).
Cis isomers obtained by rearrangement, Tuzson, Polgár and the writer (*536, 525*)

In the presence of iodine two main, crystallizable *cis* isomers, neo A and neo B appeared, both above the all-*trans* zone; yields, 16% and 23%. The respective λ_{max} shifts, 6 mμ and 5 mμ, indicate mono*cis* configurations. The *cis*-peak of neo A ($\varepsilon = 4.9 \times 10^4$) surpasses slightly the highest value observed in the α-carotene set (4.6×10^4) and almost reaches that for central-mono*cis*-β-carotene (5.2×10^4); evidently, the *cis* double bond must be located in one of the two "middle" positions (13- or 15-*cis*). Neolutein B possesses a much lower *cis*-peak and could well represent the 9- or the 9'-*cis* compound (Fig. 10, p. 32).

3,4-Dehydro-α-carotene Set.

Structure, p. 7.
Cis isomers obtained by rearrangement, Karmakar and the writer (*221*).

Upon iodine catalysis two main isomers appeared: neo U (yield, 8%; adsorbed above the all-*trans* zone; λ_{max} shift 5 mμ; practically no *cis*-peak), and neo B (18%; adsorbed below the all-*trans* form; λ_{max} shift, 10 mμ; considerable *cis*-peak). Evidently, a peripheral mono*cis* and a di*cis* isomer were formed; the latter must have a bent molecular shape.

Trollixanthin Set.

Structure, see below.
Cis isomer found in nature, Eugster and Karrer (*108*).

Both a *cis* and the all-*trans* form were isolated from plants (m. p. 143 to 145°, and 199°; λ_{max} shift, 2 mμ). It has been impossible, however, to effect a *cis* → *trans* rearrangement because the starting material was either destroyed or remained unchanged during the treatment.

(XXXVII.) Trollixanthin.

Taraxanthin Set.

Structure unknown, composition $C_{40}H_{56}O_4$; contains probably two hydroaromatic rings (trihydroxy-α-carotene monoepoxide?).
Cis isomers obtained by rearrangement, Tuzson and the writer (*536*).

This pigment yielded on catalysis by iodine the neotaraxanthins A–C, which all showed increased adsorbabilities. In CS_2, the λ_{max} shifts were, 5 mμ (neo A) and 20 mμ (neo C); a very small value was observed for neo B.

Capsanthin Set.

Structure, next page, ENTSCHEL and KARRER *(104)*; BARBER, JACKMAN, WARREN and WEEDON *(22)*; configuration, FAIGLE and KARRER *(113 a)*.

Cis isomers obtained by rearrangement, CHOLNOKY, POLGÁR and writer *(503, 367)*.

The fundamental band of this dihydroxyketone shows almost no fine structure *(Fig. 33)*. Upon addition of iodine three isomers appear. Ratio,

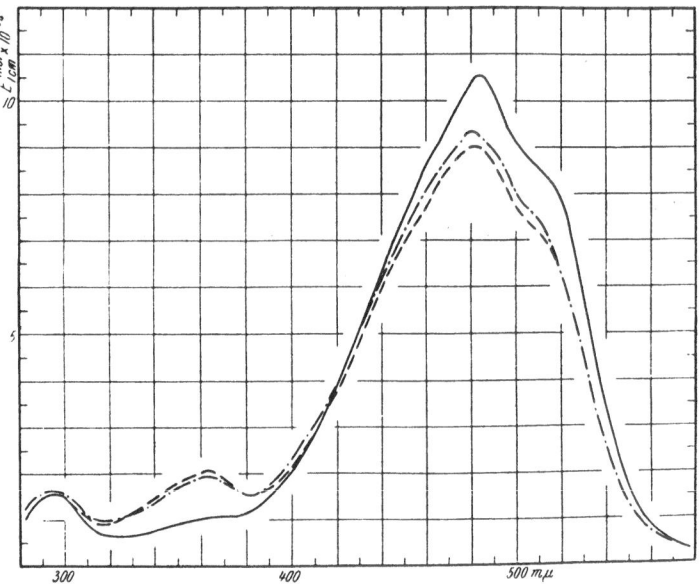

Fig. 33. Molecular extinction curves of capsanthin, in benzene: ————, fresh solution of the all-*trans* compound; ———, mixture of stereoisomers after 45 min. refluxing; and —·—·—, after catalysis by iodine *(367)*. [From: J. Amer. Chem. Soc. **66,** 186 (1944).]

all-*trans* : neo A : neo B : neo C = 66 : 23 : 7 : 4; neo A crystallized. The λ_{max} shifts observed for A and B amounted to 6 mμ (in benzene), indicating mono*cis* compounds.

Considering the respective *cis*-peak intensities ($\varepsilon = 4.4 \times 10^4$ for neo A and 2.65×10^4 for neo B), the *cis* double bond must occupy a more central position in A than in B. The minor isomer, neocapsanthin C, represents a di*cis* form (λ_{max} shift, 10.5 mμ), and its low *cis*-peak indicates an almost straight molecular shape. In chromatograms all neo forms appear below the *trans* zone.

Capsanthin-dipalmitate yielded two *cis* isomers that showed weakened adsorbabilities and did not separate sharply from each other. Ratio, all-*trans* : *cis* forms = 65 : 35.

Capsorubin Set.

Structure, see below, ENTSCHEL and KARRER *(104)*; BARBER, JACKMAN, WARREN and WEEDON *(22)*; configuration, FAIGLE and KARRER *(113 a)*.

Cis isomers obtained by rearrangement, CHOLNOKY and the writer *(503)*; CHOLNOKY et al. *(60)*.

When treated with iodine, capsorubin yields two main isomers, the neo forms A and B, adsorbed above the all-*trans* zone. Ratio, all-*trans* : *cis* forms = 70 : 30. Both neo A and B seem to be mono*cis* compounds (λ_{max} shifts, 4 mμ and 3 mμ).

Capsorubin-dipalmitate afforded two *cis* isomers that appeared below the all-*trans* zone and could not be sharply separated from one another; λ_{max} shifts, 5–6 mμ.

(XXXVIII.) Capsanthin.

(XXXIX.) Capsorubin.

2. Stereoisomeric Sets with One Hydroaromatic and One Aliphatic Terminal Group.

γ-Carotene Set.

Structure, p. 5.

Cis isomers found in nature, PINCKARD and the writer *(521, 353)*. For pro-γ-carotene, see p. 75.

Cis isomers obtained by rearrangement, POLGÁR and the writer *(526)*.

Cis isomers prepared by total synthesis, ISLER et al. *(394a)*.

Our knowledge concerning the stereochemistry of this half-aliphatic, non-symmetrical set is unsatisfactory, in part because of the great number of observed spatial forms, most of which either cannot be separated chromatographically at all or only after wasteful fractionations. The well-characterized isomer, neo-γ-carotene U (adsorbed above the all-*trans* form; λ_{max} shift, 5 mμ; low *cis*-peak) is certainly a peripheral mono*cis* compound (9- or 9'-*cis*). The *cis*-peak of the minor isomers neo H (obtained by refluxing solutions; λ_{max} shift, 5 mμ) is as high ($\varepsilon = 5.4 \times 10^4$) as that of central-mono*cis*-β-carotene (5.2 $\times$ 10^4). This isomer could possibly represent the central-mono*cis* form (15,15'-*cis*), synthesized recently by ISLER et al.; maxima in light petroleum, 347, 434, 457, 486 mμ, versus 437, 462, 494 mμ for all-*trans*-γ-carotene.

Two crystalline isomers, neo-γ-carotene P and pro-γ-carotene (p. 75) are natural products; they are absent from stereochemical equilibria.

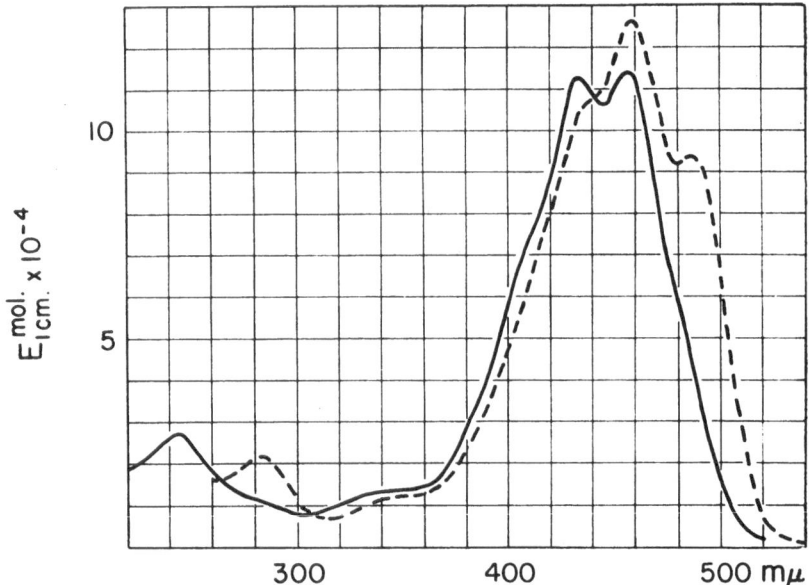

Fig. 34. Molecular extinction curves, in hexane: ———, pro-γ-carotene; and — — —, neo-γ-carotene P
(353).

Neo-γ-carotene P was isolated from the berries of *Pyracantha angustifolia*. Its spectral curve is flat in the *cis*-peak region. The small λ_{max} shift (3 mμ) would be in accordance with a mono*cis* configuration. However, IR measurements seem to indicate some unclarified feature of the neo P molecule. We have stated earlier (293) that the curves of pro-γ-carotene and neo P are almost identical; this may have been caused by some experimental error. A reinvestigation (in CS$_2$) has now revealed that there is one important difference between the two IR spectra, viz. the presence of a 13.02–13.19 μ doublet band in the neo P curve. It then seems that neo-γ-carotene P contains both methylated and unmethylated *cis* double bond(s) while the latter type is absent from pro-γ-carotene [LUNDE (290)].

It is interesting to note how profoundly this configurational difference influences the spectra in the visible and UV regions: in the fundamental band pro-γ-carotene shows two almost equal maxima (434, 457 mμ), while the neo P curve contains a single main maximum (459 mμ) (*Fig. 34*).

Gazaniaxanthin Set.

Structure unknown, possibly a monohydroxy-rubixanthin.
Cis isomers obtained by rearrangement, SCHROEDER and the writer (*532*).

As γ-carotene, gazaniaxanthin yielded on catalysis by iodine a complex chromatogram in which the *cis* zones "Group I" and "Group II" were stereochemically heterogeneous. The respective λ_{max} shifts, 5.5 mμ and 8.5 mμ, may indicate mono- and di*cis* configurations.

β-Zeacarotene Set.

Structure, 7',8'-dihydro-γ-carotene, QUACKENBUSH at al. (*352a*); ISLER at al. (*394a*).
Cis isomer prepared by total synthesis, ISLER et al. (*394a*).

The oily 15,15'-*cis* form shows a considerable *cis*-peak at 314 mμ in light petroleum. Maxima, 405, *425*, 450 mμ, versus 406, *428*, 454 mμ for the all-*trans* compound. The λ_{max} shift amounts to 3 mμ.

Celaxanthin Set.

Structure, 3(?)-hydroxy-3',4'-dehydro-γ-carotene.
Cis isomers obtained by rearrangement, LeROSEN and the writer (*283*).

When treated with iodine, this pigment yielded the neocelaxanthins A-C that were adsorbed below the all-*trans* zone; λ_{max} shifts, 3 mμ (neo A), 6.5 mμ (B), and 3.5 mμ (C). A and C must be mono*cis* isomers.

Torularhodin Set.

Structure, see below.
Cis isomer prepared by total synthesis, ISLER et al. (*196*).

Central-mono*cis*-torularhodin methylester showed a λ_{max} shift of 2 mμ.

(XL.) Torularhodin.　　COOH

3. Stereoisomeric Sets with Two Aliphatic Terminal Groups.

Lycopene Set.

Structure, p. 5.
Cis isomers obtained by rearrangement, TUZSON et al. (*536*); LeROSEN, SCHROEDER, POLGÁR, PAULING and the writer (*515*).
Cis isomers found in nature, Tables 13–15 (pp. 66, 72, 73).
Cis isomers prepared by total synthesis, ISLER et al. (*197*); GARBERS and KARRER (*128*).

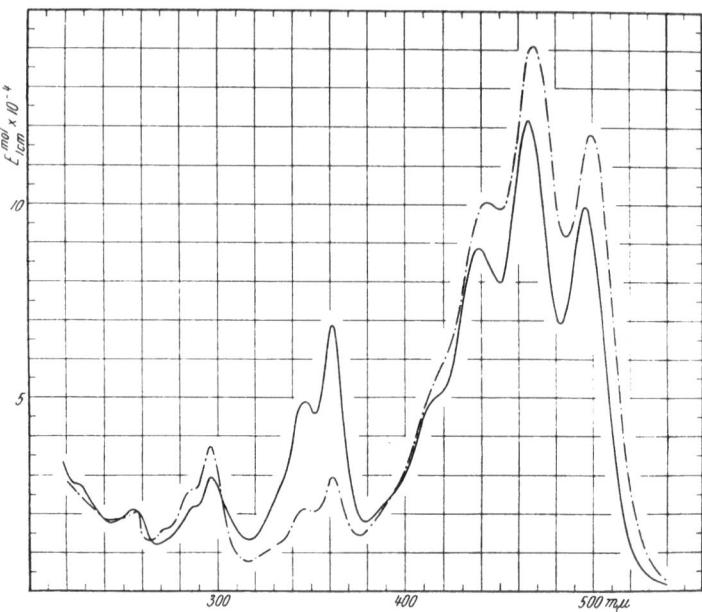

Fig. 35. Molecular extinction curves of neolycopene A: —————, fresh solution; and — · — · —, mixture of stereoisomers after catalysis by iodine (*515*). [From: J. Amer. Chem. Soc, 65, 1940 (1943).]

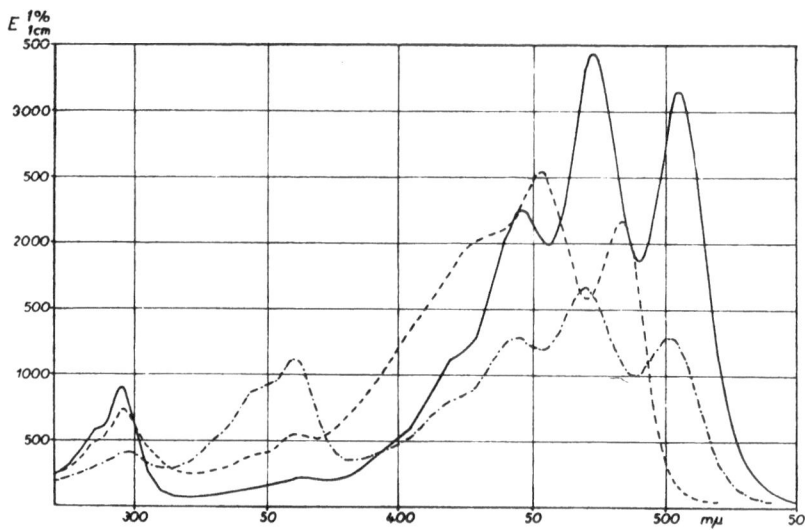

Fig. 36. Specific extinction curves, in light petroleum: — · — · —, synthetic central-mono*cis*-lycopene; —————, all-*trans*-lycopene; and — — —, 15,15′-dehydrolycopene; according to Isler et al. (*197*). [From: Helv. Chim. Acta 39, 463 (1956).]

This set occupies a unique position because about forty of its members are known (Table 15, p. 73). In contrast, only a single main isomer, neolycopene A (yield, 40–50%) appears upon rearrangement of the all-*trans* form. Considering the λ_{max} shift of 5 mμ and the extraordinarily high *cis*-peak *(Fig. 35)*, we had first assigned the central-mono*cis* configuration (15-*cis*) to this isomer. Eventually, IR readings have shown

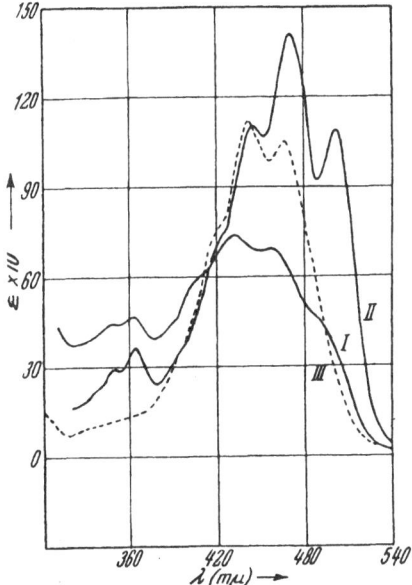

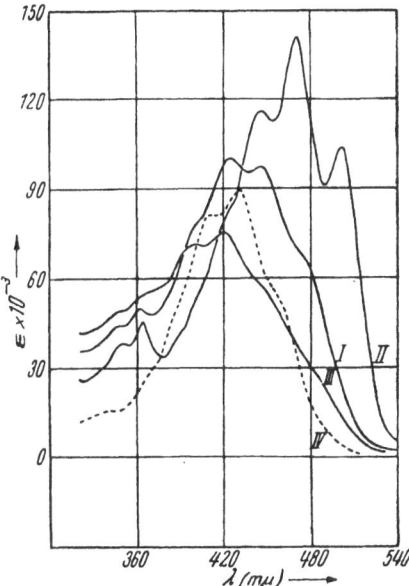

Fig. 37. Molecular extinction curves, in cyclohexane: Curve I, synthetic (hindered) *cis*-lycopene b; Curve II, the same, after catalysis by iodine; and Curve III, natural poly*cis*-lycopene "III", in hexane (*521*); according to GARBERS and KARRER (*128*). [From: Helv. Chim. Acta 36, 828 (1953).]

Fig. 38. Molecular extinction curves, in cyclohexane: Curve I, synthetic (hindered) *cis*-lycopene c; Curve II, the same, after catalysis by iodine; Curve III, synthetic (hindered) *cis*-lycopene c'; and Curve IV, natural poly*cis*-lycopene "V" in hexane (*521*); according to GARBERS and KARRER (*128*). [From: Helv. Chim. Acta 36, 828 (1953).]

that the 7.25 μ band, indicative of methylated *cis* double bonds, is present with considerable intensity. Hence, the assignment had to be changed to the next-to-central-mono*cis* configuration (13-*cis*) (*293*).

The correctness of this revision was confirmed by the synthesis of the true central-mono*cis*-lycopene (ISLER; KARRER) *(Fig. 36)*. This crystalline compound is clearly different from neolycopene A which can be obtained in solutions only. The *cis*-peak intensities are almost identical*. The Q value (p. 77) is 1.5 for the synthetic product but 1.8 for neolycopene A.

* $\varepsilon = 6.0 \times 10^4$ for the synthetic product; our value, determined for neo A indirectly, is evidently too high (6.8 $\times$ 10^4) (*515*).

7*

The minor isomer, neolycopene B, shows a λ_{max} shift of 8 mμ and a *cis*-peak of medium intensity. It seems to possess two *cis* double bonds, one of which has the same location as that of neolycopene A. Indeed, some neo B is spontaneously formed from neo A when the solution is kept at room temperature.

Three hindered members of this set, the *cis*-lycopenes b, c, and c', were synthesized by KARRER et al. (*128*). Although they are relatively labile and not crystallizable, their preparation has proved that hindered *cis* lycopenes do exist and are characterized by degraded spectra. These compounds show *cis*-peaks (with fine structure) and must possess bent molecular form (*Figs. 37–38*, p. 99). Their *cis*-peaks differentiate them from our poly*cis* lycopenes ex *Pyracantha* (cf. p. 69).

Poly*cis*-lycopenes were discussed on pp. 67–74. The IR curve of prolycopene shows, with high intensities, the presence of both methylated and unmethylated *cis* double bonds. The poly*cis* character of prolycopene was confirmed in stepwise isomerization experiments (p. 70).

Neurosporene Set.

Structure, 7,8–dihydrolycopene (p. 102) (*380, 87;* cf. also *109*)
Cis isomer found in nature, p. 74.
Cis isomers obtained by rearrangement, MAGOON and the writer (*300*).

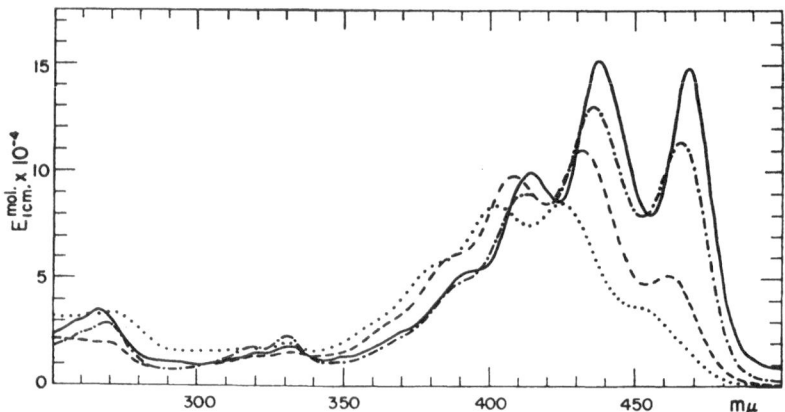

Fig. 39. Molecular extinction curves of stereoisomeric neurosporenes, in hexane: ———, all-*trans*; —— —, neo P (proneurosporene); · · · · ·, neo R; and —·—·—, iodine-catalyzed equilibrium mixture (*300*). [From: Arch. Biochem. Biophys. 68, 263 (1957).]

Upon treatment of all-*trans*-neurosporene (ex *Neurospora crassa* or neo P), the six neo forms, A–F, were obtained, A–D in remarkably similar quantities. Ratios, unchanged all-*trans* : neo A : neo B : neo C : : neo D : neo E : neo F = 44 : 13 : 11 : 12 : 15 : 3 : 2. Although no con-figurational assignments can be made at the present time, mono*cis* forms are clearly preponderant in this set, (λ_{max} shifts in the above sequence: 0, 1, 5, 5.5, 7, 9, and 14 mμ) *(Fig. 39).* For neo R see also p. 75.

ζ-Carotene Set.

Structure, 7, 8, 7′, 8′-tetrahydrolycopene *(380, 87).*
Cis isomer obtained by total synthesis, DAVIS, JACKMAN, SIDDONS and WEEDON *(87).*

Synthesis has yielded both all-*trans*-ζ-carotene (380, *401,* 425 mμ, in hexane) and the central-mono*cis* form (378, *398,* 422 mμ); the latter is perhaps also a natural product.

Phytofluene Set.

Structure, see next page: 7,8, 11,12, 8′,7′-hexahydrolycopene *(499, 380, 87).*
Cis isomers occurring in nature and obtained by rearrangement, PETRACEK, KOE and the writer *(350, 250);* WALLACE and PORTER *(474).*

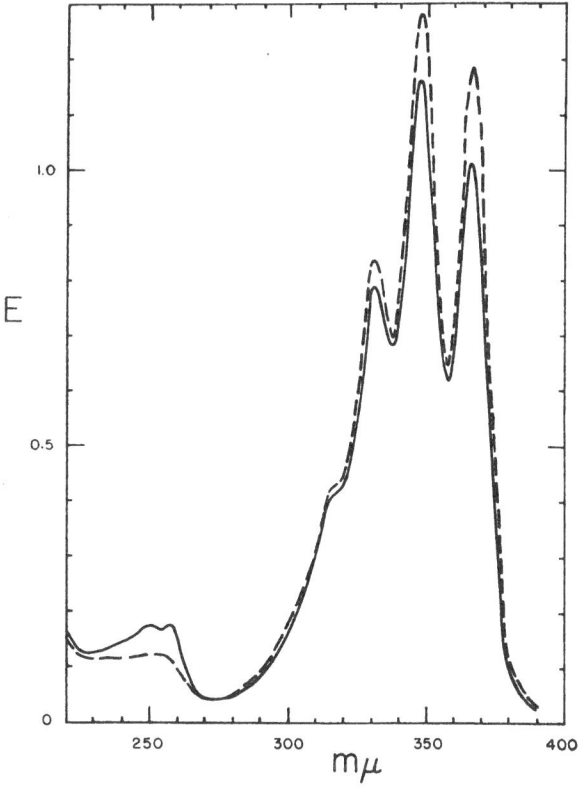

Fig. 40. Extinction curves of tomato phytofluene (*cis*), in hexane: ————, fresh solution; and — — —, after catalysis by iodine (*350*). [From: J. Amer. Chem. Soc. 74, 184 (1952).]

In this nearly colorless set which contains only five conjugated double bonds, stereochemical work is difficult because of the oily nature and photosensitivity of the substance. The main phytofluene form in the tomatoes studied by our group was a mono*cis* phytofluene (λ_{max} shift, 0.5–1 mμ) *(Fig. 40*, p. 101*)*. Considering the low *cis*-peak, the *cis* double bond cannot be located at the center of the phytofluene chromophore (position 13'); the hindered 11'-position is excluded by the spectrum. A 15'-*cis* configuration would possibly account for all known facts. The 13-*cis* configuration cannot be excluded.

(XLI.) Neurosporene.

(XLII.) Phytofluene.

(XLIII.) Phytoene.

Phytoene Set.

Structure, 7,8, 11,12, 12',11', 8',7'-octahydrolycopene *(379, 87)*.
Cis isomer found in nature (?), RABOURN and QUACKENBUSH *(379)*.
Cis isomer obtained by total synthesis, WEEDON et al. *(87)*.

Phytoene ex carrot oil is possibly a *cis* compound but it could not be isomerized either by heat or by iodine. WEEDON has obtained a stereo-isomeric mixture, practically identical in its spectrum with natural phytoene *(275, 286, 298* mμ, in hexane); it was partially resolved via the thiourea complex.

Rhodoviolascin (Spirilloxanthin) Set.

Structure, p. 7 *(23; 214;* cf. *224)*.
Cis isomers obtained by rearrangement, POLGÁR, VAN NIEL and the writer *(364)*.

The spectral curve of this dimethoxy compound appears in *Fig. 41*. The sole main *cis* isomer in this set is neo A which is formed easily in about 40% yield. The all-*trans* configuration of natural rhodoviolascin is remarkably labile and even fresh solutions of the crystals contain appreciable amounts of neo A. The λ_{max} shift (7 mμ in hexane or 10.5 mμ in benzene) seems to indicate a di*cis* compound whose molecules must possess a bent shape ($\varepsilon = 5.1 \times 10^4$ at *cis*-peak). It is believed that one of the *cis* double bonds has the central or next-to-central location.

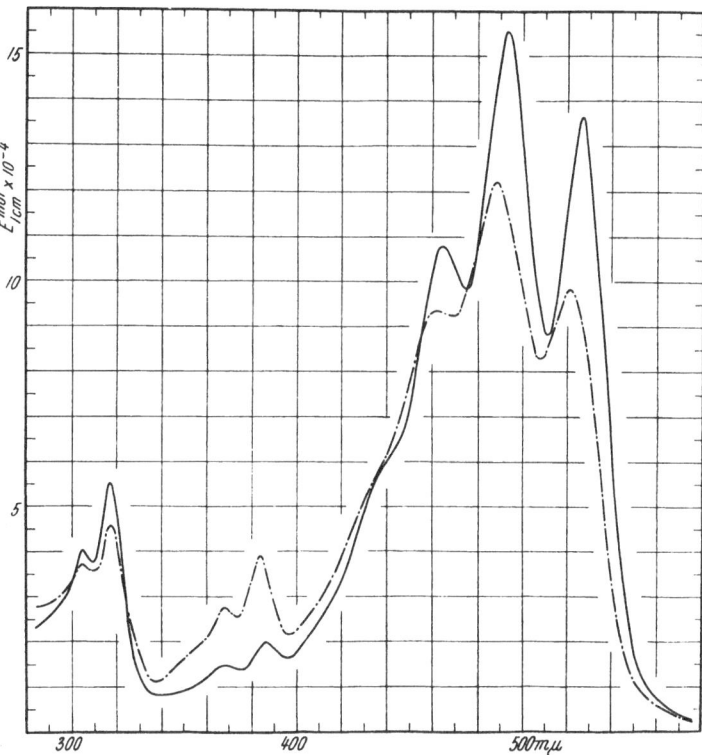

Fig. 41. Molecular extinction curves of rhodoviolascin, in hexane: ————, fresh solution of the all-*trans* compound; and — · — · —, mixture of stereoisomers after catalysis by iodine (*364*). [From: Arch. Biochemistry 5, 243 (1944).]

Fucoxanthin Set.

Structure unknown, probably fully aliphatic; composition, $C_{40}H_{56}O_6$.

Cis isomers found in nature and obtained by rearrangement, STRAIN, MANNING and HARDIN (*424, 426*).

This algal and diatom pigment yields two isomers, neo A and neo B which are adsorbed above the all-*trans* form (λ_{max} shifts in ethanol, 6–7 mμ). The same stereoisomeric mixture (all-*trans* : *cis* forms = 90 : 10) is obtained when diatoms are rapidly extracted. The thermal interconversion of the fucoxanthins is accelerated by light and by contacting petroleum ether solutions with sugar or powdered glass which strongly retain the pigment. Considering the HI-sensitivity, iodine catalysis is successful only in the presence of some organic base.

4. Stereoisomeric Sets with Two Aromatic Terminal Groups.

1,18-Diphenyl-3,7,12,16-tetramethyl-octadecanonaene Set.

Structure, see below.

Cis isomers prepared by total synthesis, GARBERS, EUGSTER and KARRER (*106, 125, 126*).

Several *cis* members of this set were synthesized by KARRER's research group. Although some hindered and unhindered *cis* forms could not

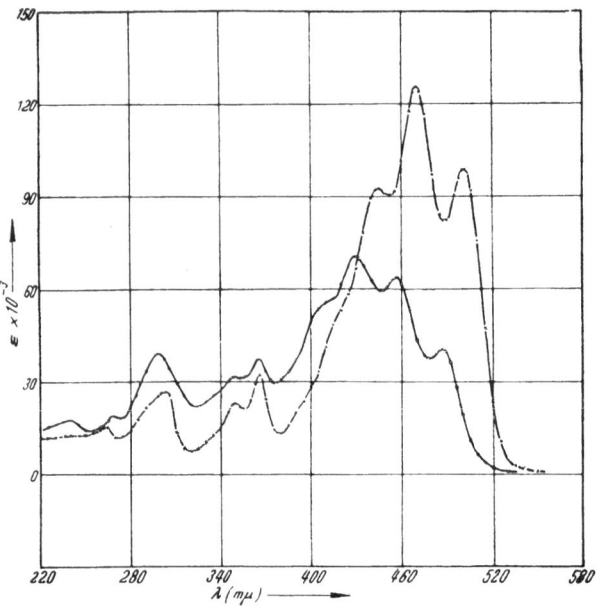

Fig. 42. Molecular extinction curves of synthetic (hindered) 5-*cis*-1,18-diphenyl-3,7,12.16-tetramethyl-octadecanonaene, in cyclohexane: —o—o—, fresh solution; and — × — × —, mixture of stereoisomers after catalysis by iodine (higher curve); according to GARBERS, EUGSTER and KARRER (*126*). [From: Helv. Chim. Acta 36, 562 (1953).]

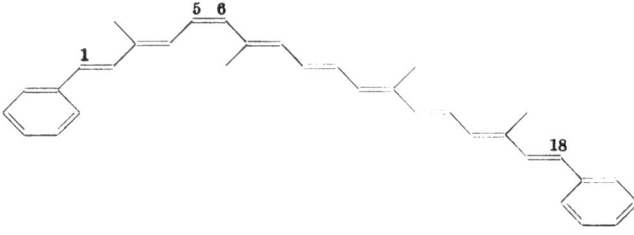

(XLIV.) 5-*cis*-1,18-Diphenyl-3,7,12,16-tetramethyl-octadecanonaene (the *cis* double bond is sterically hindered) (cf. *Fig. 42*).

(XLV.) 5,13-Dicis form of the same compound (both cis double bonds are sterically hindered).

yet be assigned configurations, the existence and stability of the well-defined, crystalline, hindered members (XLIV) and (XLV) have theoretical importance. They show degraded spectra.

Renieratene and Isorenieratene Sets.

Structures, see below.

Mixture of cis isomers obtained by rearrangement, YAMAGUCHI (487–489).

These sponge pigments are closely related to the synthetic polyenes just mentioned. Renieratene and isorenieratene show upon isomerization with iodine the expected spectral changes, including *cis*-peaks. No individual *cis* forms have been described up to the present time.

(XLVI.) Renieratene.

(XLVII.) Isorenieratene.

1,3,7,12,16,18-Hexaphenyl-octadecanonaene Set.

Structure, see next page.

Cis isomers prepared by total synthesis and by rearrangement, ZIEGLER, EUGSTER and KARRER (540).

Four crystalline, hindered *cis* members of this set, the "*cis* forms I–IV" are known. They have degraded spectra and are iodine-sensitive, yielding, besides the all-*trans* form, a number of new *cis* isomers whose curves indicate the presence of unhindered *cis* double bonds. One of the isomerization products might well represent a second all-*trans* form that differs from the ordinary one by a crosswise orientation of phenyl side-chains (*540*).

$$\left[\underset{}{\bigcirc}\!\!-\text{CH}=\text{CH}-\underset{\underset{\bigcirc}{|}}{\text{C}}=\text{CH}-\text{CH}=\text{CH}-\underset{\underset{\bigcirc}{|}}{\text{C}}=\text{CH}-\text{CH}= \right]_2$$

(XLVIII.) 1,3,7,12,16,18-Hexaphenyl-octadecanonaene.

"*Naphthyl-carotene*" *Set.*

Structure, 1,18-di-β-naphthyl-3,7,12,16-tetramethyl-octadecanonaene.
Cis isomer prepared by total synthesis, LINNER, EUGSTER and KARRER (*287*).

A sterically unhindered mono*cis* form (λ_{max} shift in CS_2, 4 mμ) was isolated and showed remarkable thermostability (m. p. 225°). The extraordinarily high *cis*-peak indicates a central position of the *cis* double bond.

X. Lower-molecular Weight Carotenoid-carboxylic Acids: Bixin and Crocetin.

The carbon skeletons of these stereochemically important pigments are identical with the middle section of the β-carotene molecule but the chromophore is terminated by conjugated carboxyl groups. Hence, it can be tentatively assumed that bixin, $CH_3OOC \cdot C_{22}H_{26} \cdot COOH$, and crocetin, $HOOC \cdot C_{18}H_{22} \cdot COOH$, are products of the bio-oxidative cleavage of primarily formed C_{40}-carotenoids.

Bixin Set.

This monomethylester of a nonaene-dicarboxylic acid (IL) occurs in the seed hulls of *Bixa orellana* in substantial amounts and constitutes the main pigment of industrial Orlean. The dimethylester is termed methylbixin and the free acid, norbixin. When methylated, the bixin molecule acquires symmetry whereby the calculated number of *cis-trans* forms decreases from 512 to 272 and that of the unhindered isomers from 32 to 20. Considering this simplification and the low solubility of bixin and norbixin, most stereochemical work has been carried out with methylbixin.

(IL.) Bixin (all-*trans* form).

Historically, bixin was the first *cis* polyene observed in nature. In 1923, HERZIG and FALTIS (*161*) in a single, unreproducible experiment have obtained from Orlean a second bixin with higher melting point and longer wavelength spectrum. Presumably, an accidental catalyst had caused a *cis* → *trans* rearrangement during isolation. Six years later, KARRER et al. (*230*) tentatively assumed that the two bixins are in the relationship of geometrical isomerism. They showed that natural bixin can be converted into HERZIG's bixin by means of iodine of which, according to KUHN and WINTERSTEIN (*270*), catalytic amounts suffice.

The structural identity of the two bixins has been demonstrated as follows [KUHN et al. (*260, 261*)].

Since both pigments had afforded the same dihydro derivative, it became possible to convert the Orlean bixin into HERZIG's bixin by reduction and subsequent air oxidation (in the presence of piperidine):

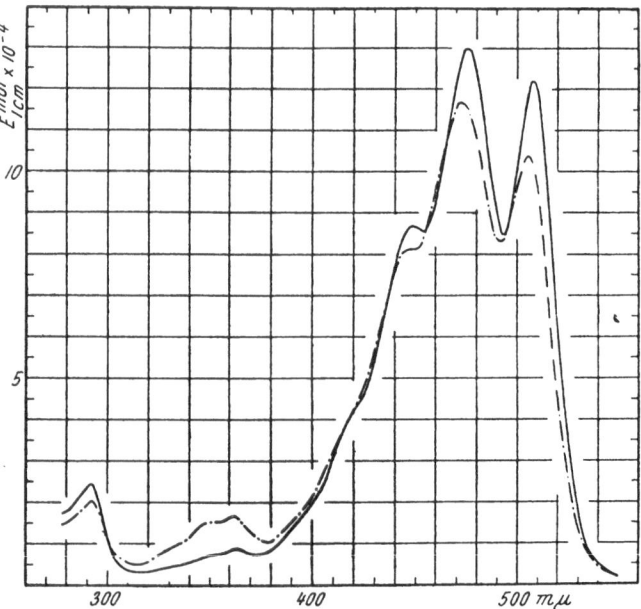

Fig. 43. Molecular extinction curves of all-*trans*-methylbixin, in benzene: ————, fresh solution; and — · — · —, mixture of stereoisomers after catalysis by iodine (*509*). [From: J. Amer. Chem. Soc. **66**, 322 (1944).]

natural bixin→dihydro-bixin→second bixin (*260, 270*). An analogous transition was realized by methylation followed by saponification: natural bixin→natural methylbixin→second bixin.

In the literature, several names have been used for the two bixins: Bixin = ordinary bixin = natural bixin = Orlean bixin = *cis*-bixin = α-bixin = labile bixin = = bixin II = lower melting bixin (m. p. 198°).—Second bixin = isobixin = *trans*-bixin = all-*trans*-bixin = β-bixin = stable bixin = bixin I = higher melting bixin (m. p. 220°).—Another "isobixin" described by VAN HASSELT (*152*) could not be reproduced by KARRER and TAKAHASHI (*239*); its homogeneity is doubtful.

Turning to stabilities, we may stress that as in the class of the C_{40}-carotenoids the simple terms "stable" and "labile" are unsatisfactory. The natural product, "labile" bixin, shows a high degree of thermo-stability and is also relatively photostable but is extremely sensitive to iodine, in light. We will designate the main pigment of the *Bixa* seeds as "natural bixin" and the dimethyl ester as "natural methyl-bixin", although the latter does not seem to occur in plants. The pigment obtained by iodine catalysis is all-*trans*-bixin.

Strictly speaking, the name "natural bixin" is no longer a precise term because small quantities of the all-*trans* form have been isolated from *Aristolochia cymbifera* roots under conditions which exclude *cis-trans* rearrangement (*144*).

Since earlier studies had been restricted to the two bixins mentioned and since the reversibility of the stereoisomerization had not been clearly

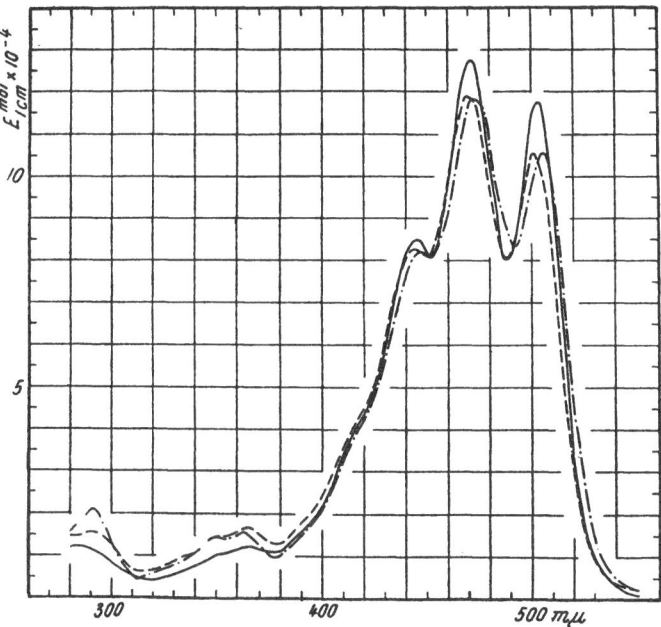

Fig. 44. Molecular extinction curves of natural methylbixin, in benzene: ————, fresh solution; — — —, mixture of stereoisomers after refluxing in darkness for 45 min.; and — · — · —, after catalysis by iodine (*509*). [From: J. Amer. Chem. Soc. 66, 322 (1944).]

claimed, ESCUE and the writer (*507*, *509*) have re-investigated the bixin set. Upon spatial rearrangement three new *cis* forms, the neomethyl-bixins A, B, and C were isolated, A and C in crystalline form *(Figs. 43–46)*. A few years ago, totally-synthetic all-*trans*-methylbixin was prepared by three research groups [AHMAD and WEEDON (*2*), INHOFFEN and RASPÉ (*190*), ISLER et al. (*198*)] and identified with the pigment that can be obtained from "natural" methylbixin by means of iodine. INHOFFEN also described the crystalline central-mono*cis*-methylbixin *(Fig. 47*, p. 112) and found it different from the neo compounds A–C. Its configuration is confirmed

by the IR spectrum, viz. absence of methylated *cis* double bonds (*293*).
Central-mono*cis*-methylbixin does not occur in stereoisomeric equilibria,
pending experiments to be conducted in red light.

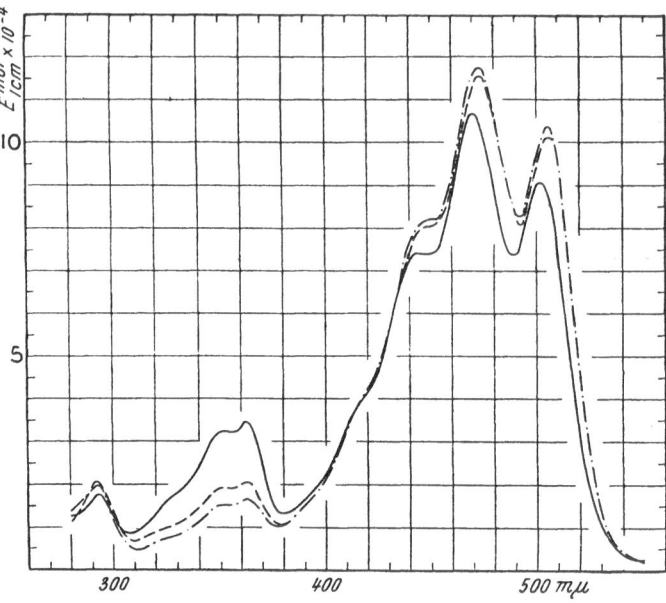

Fig. 45. Molecular extinction curves of neomethylbixin A, in benzene: ————, fresh solution; — — —,
mixture of stereoisomers after refluxing in darkness for 45 min.; and — · — · —, after catalysis by iodine (*509*).
[From: J. Amer. Chem. Soc. **66**, 322 (1944).]

Some characteristics of the five known crystalline members of the
methylbixin set appear in *Table 17*.

Table 17. Stereoisomeric Methylbixins (cf. *509*, *190*).

| Name | Adsorbed above (a) or below (b) the all-*trans* form | Melting point | Position of λ_{max} | | λ_{max} shift (mμ) | ε at *cis*-peak | | Q value (p. 77) |
			in benzene (mμ)	in petr. ether (mμ)		in benzene	in petr. ether	
All-*trans*.	—	220°	475	458	—	—	—	—
Natural .	a	198°	471	—	4	1.2×10^4	—	10.7
Neo A ..	b	190°	470	—	5	3.5×10^4	—	3.1
Neo C ..	b	150–151°	465	—	10	2.2×10^4	—	4.4
Central-mono*cis*	b	193–194°	—	455	3	—	6.1×10^4	1.6

Considering the λ_{max} shift values, neo C represents a di*cis* form but
both natural methylbixin and neo A must be mono*cis* compounds. On

the basis of the high *cis*-peak we had assumed earlier that the *cis* double bond is located in the center of the neo A molecule. Since, however, the intensity of this peak is surpassed by that of INHOFFEN's synthetic product, neomethylbixin A is now assigned the next-to-central-mono*cis* configuration (8-*cis*).

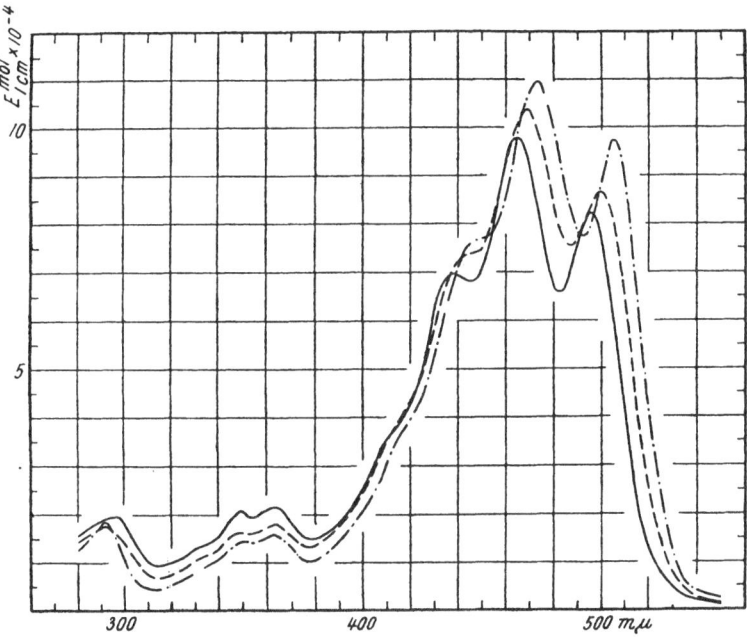

Fig. 46. Molecular extinction curves of neomethylbixin C, in benzene: ————, fresh solution; — — —, mixture of stereoisomers after refluxing in darkness for 45 min.; and — · — · —, after catalysis by iodine (509). [From: J. Amer. Chem. Soc. 66, 322 (1944).]

The reported absence of bands both at 7.25 μ and 12.0–14.0 μ remains unexplained.

This leaves the positions 2, 4, and 6 open for the *cis* double bond in natural methylbixin. Although the biochemically interesting configurational problem of natural methylbixin can still not be solved, as we will see a tentative assignment can be made.

In the *cis*-peak region, the spectral curve of this isomer is almost flat as foreseen for a peripheral mono*cis* compound. A configuration of this "neo U type" is further indicated by the strong adsorption affinity that surpasses that of the all-*trans* pigment.

Natural methylbixin is not a component of the *cis-trans* equilibrium mixtures that are obtained from all-*trans*-methylbixin by refluxing, insolation, or iodine catalysis: in these instances only the neo forms A–C appear. Neither is all-*trans* pigment formed in any marked amounts when natural methylbixin is insolated or refluxed. In the absence of iodine the thermo- and photostability of natural methylbixin is so high that the molecule rather undergoes further *trans* → *cis* rearrangement than to rotate its *cis* double bond into the *trans* configuration. Iodine

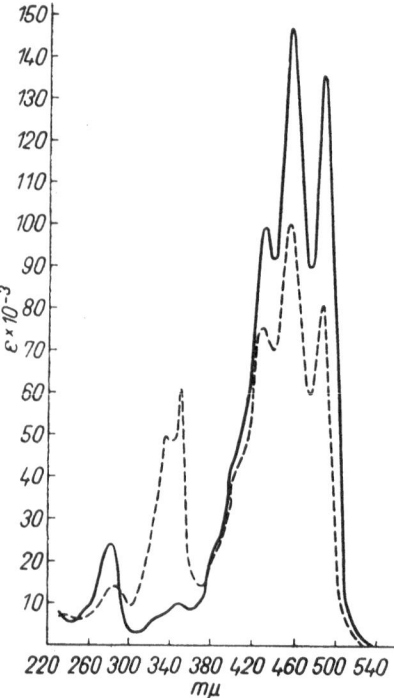

Fig. 47. Molecular extinction curves of methyl-bixin, in light petroleum: —————, all-*trans* form; and — — —, central-mono*cis* form; according to Inhoffen and Raspé (*190*). [From: Liebigs Ann. Chem. **592**, 214 (1955).]

Fig. 48 (right). Stereochemically important sections taken from infrared curves of some *cis-trans* isomeric methylbixins: saturated CCl₄ solutions (1 mm. cell) in the 7.0—7.5 μ and 8.0—9.0 μ regions; mineral oil mulls in the 10.0—10.5 μ and 12.0—14.0 μ regions (*293*). [From: J. Amer. Chem. Soc. **77**, 1647 (1955).]

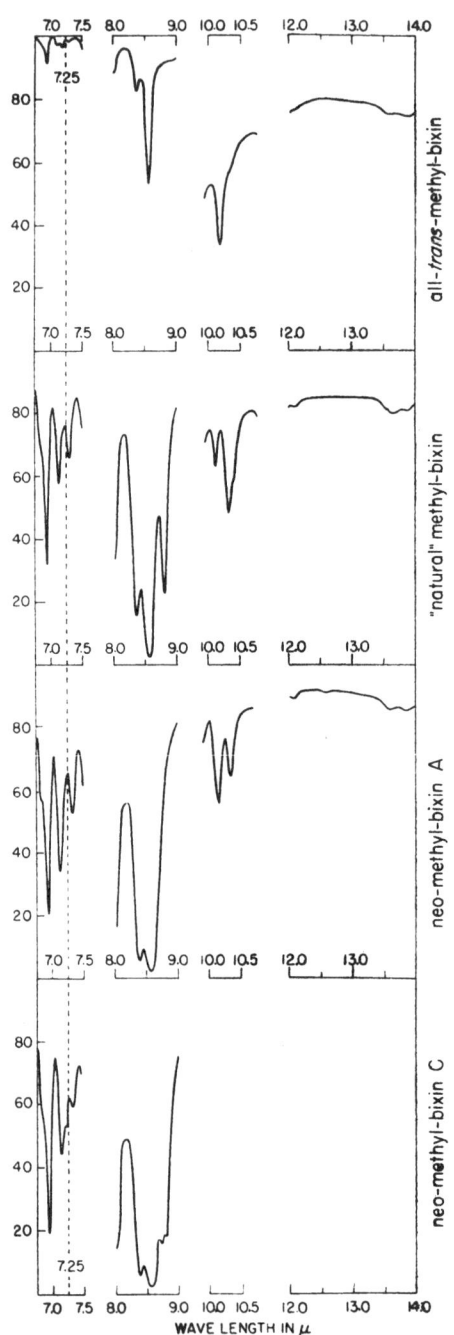

brings about the expected equilibrium mixture where from natural methyl-bixin has disappeared. In contrast, the interconversion of natural methylbixin and the di*cis* isomer neo C is easy. Therefore the location of one of the *cis* double bonds in neo C is probably identical with that of the natural pigment. At present the sole in vitro source of natural methylbixin is neo C (ex all-*trans*) that yields it in substantial amounts upon various treatments (in the absence of iodine).

If full arrows stand for easy interconversion and dotted arrows for difficult or impossible ones, then we may write the following scheme, assuming that iodine is absent:

All-*trans*-methylbixin ⇄ Neomethylbixin A
(mono*cis*)

Natural methylbixin ⇄ Neomethylbixin C
(mono*cis*)   (di*cis*)

Typical examples of equilibrium mixtures (*509*):

All-*trans*-methylbixin: Refluxing; unchanged *trans* : neo A : minor isomers = = 63 : 35 : 2.—Insolation; unchanged *trans* : neo A = 94 : 6.—Iodine, in daylight; unchanged *trans* : neo A : other neo forms = 72 : 19 : 9.

Natural methylbixin: Refluxing; unchanged natural form : neo C (+ minor isomers) = 74 : 26.—Insolation; unchanged natural form : minor isomers : neo C = = 90 : 2 : 8.

Neomethylbixin C: Refluxing; natural form : all-*trans* : neo A : neo B : unchanged neo C = 33 : 12 : 6 : 3 : 46.—Insolation; natural form : neo A : unchanged neo C = = 21 : 2 : 77.

The peculiar behavior of natural methylbixin can be tentatively explained by the influence of a conjugated carbonyl group on the *cis* double bond, i. e. by the 2-*cis* configuration. Although this assignment is in accordance with the flatness of the curve in the *cis*-peak region, it necessitates further discussion because the 2-position is a hindered one and the natural methylbixin curve is not degraded. This situation might be explained by the special character of the $CH_3OOC—CH=CH—$ group which could assume a conformation essentially *normal* with reference to the plane of the rest of the molecule.

The proposed 2-*cis* configuration is in accordance with the IR spectrum (*Fig. 48*, p. 112) (*293*): (a) The 7.25 μ band and hence a methylated *cis* double bond is absent; (b) a strong band appears at 8.00 μ that does not occur in the all-*trans* spectrum. It may well represent a C—O stretching vibration for which a conjugated *cis* double bond is responsible. A more remote such bond could not have this effect.

Our configurational assignment to natural methylbixin, and hence to bixin itself, is not in accordance with KARRER and SOLMSSEN's earlier observations (*238*) who submitted the all-*trans* and natural bixins to a parallel treatment with per-manganate. On the basis of the identity of some of the aldehydes obtained from both sources and the steric difference between those two aldehydes that were obtained by cleavage at the C=C double bond third from the free carboxyl group, these authors tentatively concluded that bixin was the 6-*cis* compound. This interesting approach, which would require re-investigation, does show that the *cis* double bond cannot be located in the middle section of the natural bixin molecule.

If natural methylbixin is the proposed 2-*cis* form, then neo C must have one of the following eight di*cis* configurations: 2,4; 2,6; 2,8; 2,10; 2,8'; 2,6'; 2,4'; or 2,2'. Of these 2,6 and 2,6' are eliminated because of steric hindrance (the spectrum is not degraded); furthermore, 2,4; 2,4'; and 2,2' are excluded because they would contradict the marked *cis*-peak. Consequently, neo C may be either 2,8- or 2,10- or 2,8'-di*cis*-methylbixin.

The reader will note that the foregoing conclusions have tentative character and would require confirmation. Recently, on the basis of nuclear magnetic resonance readings, BARBER, JACKMAN and WEEDON (*24*) have advanced strong arguments for the 4-*cis* configuration of "natural" methylbixin.

Crocetin Set.

Crocetin (L), a lower homolog of bixin, is the principal pigment of saffron *(Crocus sativus)* in which it occurs mainly in the form of the di-gentiobioseester crocin. Crocetin has also been observed in several other flowers and in some fruits (cf. *232*).

(L.) Crocetin.

Crocetin is a heptaene-dicarboxylic acid. Its methyl- and dimethyl-ester are termed, respectively, methylcrocetin and dimethylcrocetin.

An earlier nomenclature (crocetin = α-crocetin; monomethylester = β-crocetin; and dimethylester = γ-crocetin) has been abandoned.

The calculated number of stereoisomeric crocetins is 72, and that of the unhindered forms is 20. In contrast to the bixin molecule, the double bonds adjacent to the carbonyls are sterically unhindered. Only three members of this set are known. The all-*trans* and a *cis* form originate from plants while the central-mono*cis* isomer is a product of total synthesis; it can be rearranged to all-*trans*-crocetin. As in the bixin set, most of the stereochemical work has been carried out with the relatively easily soluble dimethylester.

KUHN and WINTERSTEIN (*271*, *272*) should be credited with the first observation about geometrical isomerism in the crocetin set. On trans-esterification of the native glucoside (from Spanish saffron) they isolated, besides the well-known all-*trans*-dimethylcrocetin (m. p. 222°) a second pigment (m. p. 141°) that was interpreted as a *cis* compound on the basis of the lower melting point, higher solubility, spectrum, and relatively greater lability. Its quantities in the stigmata are much inferior to those of the all-*trans* compound; and subsequently, the attempts to obtain *cis*-crocetin from various other saffron samples have failed altogether. Recently, it was possible, however, to repeat the isolation (cf. *182*). Perhaps working in red light would improve the yields.

The following names have been used for the two natural dimethylcrocetins (and analogous ones for the corresponding crocetins): Stable dimethylcrocetin = = *trans*-dimethylcrocetin = all-*trans*-dimethylcrocetin = dimethylcrocetin I = higher melting dimethylcrocetin (m. p. 222°).—Labile dimethylcrocetin = *cis*-dimethylcrocetin = natural labile dimethylcrocetin = natural *cis*-dimethylcrocetin = lower melting dimethylcrocetin (m. p. 141°).

The steric nature of the relationship between these two pigments was demonstrated also by chemical conversions: (a) Since the reduction of both dimethylcrocetins afforded the same dihydro compound, the following transition was realized (*260*): natural *cis*-dimethylcrocetin → dihydro derivative → all-*trans*-dimethylcrocetin. (b) Saponification of the *cis* ester in the heat has furnished all-*trans*-crocetin.

According to KUHN and WINTERSTEIN (*271*) direct *cis* → *trans* rearrangement takes place upon melting crystals or adding to solutions trace amounts of iodine in diffuse daylight. The most impressive method is, however, illumination in the absence of catalysts. Although the configuration did not change when *cis* crystals were kept in darkness for 20 years (*263*), during the observation of a fresh solution in the visual spectroscope the bands migrated towards longer waves and soon reached the positions characteristic for all-*trans*-dimethylcrocetin. Under certain conditions of artificial illumination the observed half-time of the rearrangement was 6 minutes. The in vitro photosensitivity of natural *cis*-dimethylcrocetin by far surpasses that of the two other isomers.

The main spectral maxima of all-*trans*-dimethylcrocetin (in light petroleum) were found at 422, *448* mμ and at 422, *450* mμ (*188*, *198*). KUHN and WINTERSTEIN (*271*) had reported 420, *445.5* mμ for the all-*trans* compound and 416, *442* mμ for the *cis* form ex saffron. The small λ_{max} shift indicates a mono*cis* configuration. A priori four locations can be considered for the *cis* double bond in natural *cis* dimethylcrocetin, namely the positions 2, 4, 6, and 8. Of these the non-degraded spectrum eliminates position 4 and the moderate *cis*-peak excludes the central position 8. Because of the new feature of extreme photosensitiveness, which could well be caused by a carbonyl group in conjugation with the *cis* C=C double bond, it seems reasonable to give preference to the 2-*cis* configuration at the present time, pending confirmation by synthesis.

8*

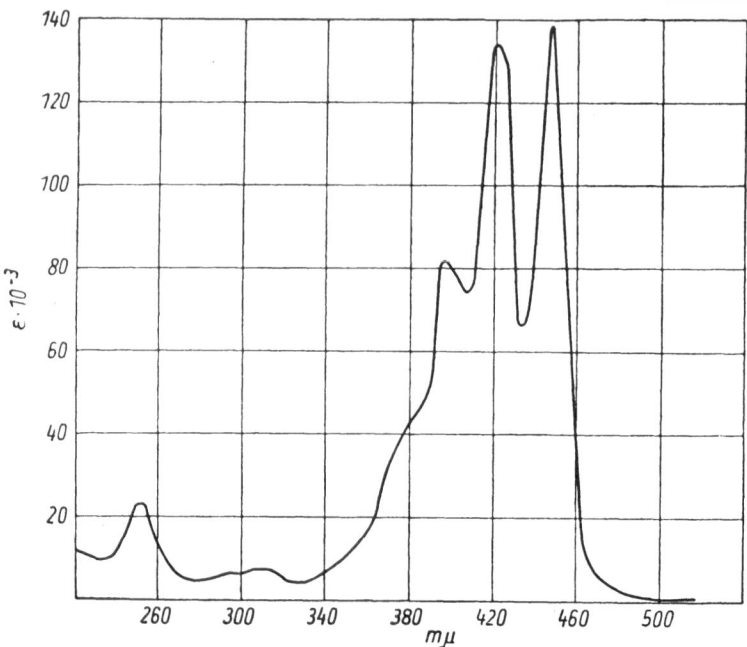

Fig. 49. Molecular extinction curve of all-*trans*-dimethylcrocetin, in light petroleum; according to INHOFFEN et al. (*188*). [From: Liebigs Ann. Chem. 580, 7 (1953).]

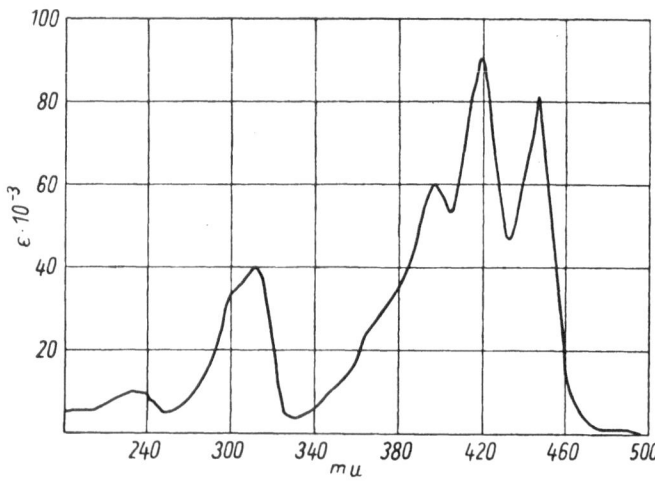

Fig. 50. Molecular extinction curve of central-mono*cis*-dimethylcrocetin, in light petroleum; according to INHOFFEN et al. (*188*). [From: Liebigs Ann. Chem. 580, 7 (1953).]

The stability difference between 2(?)-*cis*-dimethylcrocetin and 2(?)-*cis*-methyl-bixin might be caused by the respective positions of the terminal *cis* double bonds: —CH=C(CH₃)COOH and —CH=CH—COOH.

Other authors have considered the 6-*cis* configuration (*263*).

All-*trans*-dimethylcrocetin (m. p. 146°) (*Fig. 49*, p. 116) was obtained by total synthesis and identified with the main saffron pigment [INHOFFEN et al. (*188*); ISLER et al. (*197, 198*)]. The last intermediate in the INHOFFEN synthesis was central-mono*cis*-dimethylcrocetin (also termed 8-*cis* or 8,8'-*cis*) (*Fig. 50*). Its main maxima lie at *420, 446* mµ; and the value of the λ_max shift (2 mµ) is smaller at the center of the chromophore than in the peripheries also in this relatively short conjugated system (cf. p. 78).

Comparative infrared measurements in the dimethylcrocetin set were reported by KUHN, INHOFFEN, STAAB and OTTING (*263*) and confirmed by LUNDE and the writer (*293*) (*Fig. 51*). The results agree with the conclusions given above. Among the mono*cis*-dimethyl-crocetins only the central *cis* form has a symmetrical configuration. Accordingly, the IR curve of the *cis* compound (ex saffron) shows doublets or triplets at several wavelengths where singlets occur in the curve of the synthetic *cis* form. Furthermore, in case of the central-mono*cis*-dimethylcrocetin an intense band appears at 12.88 µ representing an unmethylated *cis* double bond (cf. p. 79). Because this maximum is missing in the curve of the natural *cis* isomer, the latter cannot have a *cis* configuration about the 4- or 8- double bond. As expected the 7.25 µ band, characteristic of methylated *cis* double bonds, is absent from the all-*trans* and central-mono*cis* spectra.

The properties of the synthetic and natural *cis* dimethylcrocetins were carefully compared by KUHN, INHOFFEN, STAAB and OTTING (*263*): Both substances show the phenomenon of double melting points (p. 19); the fundamental bands of the two isomers are very similar; the synthetic *cis* form is distinguished from the natural one (or from all-*trans*-dimethylcrocetin) by its higher *cis*-peak, the IR spectrum, and the Debye-Scherrer diagram.

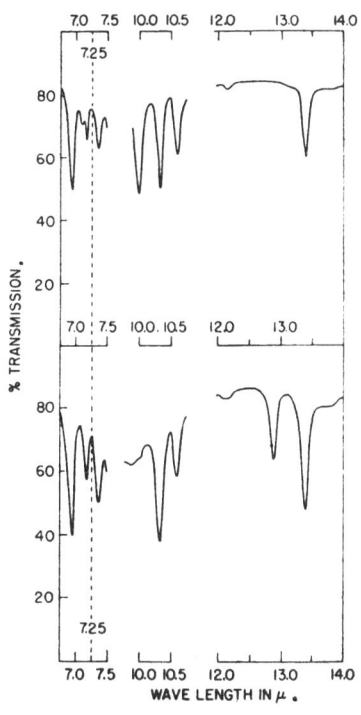

Fig. 51. Stereochemically important sections taken from infrared curves of all-*trans*-dimethylcrocetin (top) and central-mono*cis*-dimethylcrocetin (bottom): saturated CCl₄ solutions (1 mm. cell) in the 7.0–7.5 µ and 10.0–10.5 µ regions; mineral oil mulls in the 12.0–14.0 µ region (*293*). [From: J. Amer. Chem. Soc. 77, 1647 (1955).]

XI. *Cis-trans* Isomerism and Provitamin A Effect of Carotenoids.

The chemical and biological relationships between carotene and vitamin A have been established by the classical investigations of STEENBOCK (*412*), MOORE (*320*), EULER (*111*), KARRER (*112*), and others. It is now well known that β-carotene whose molecules contain two unsubstituted β-ionone rings is partially converted in the body into vitamin A, according to the schematic expression, $C_{40}H_{56} \rightarrow C_{20}H_{29}OH$. Several other naturally occurring carotenoids such as α-carotene, γ-carotene, and cryptoxanthin undergo analogous cleavage but in these instances only one half of each molecule may theoretically yield the bio-active substance. Although the postulate that at least one unsubstituted β-ionone ring must be present in provitamins A has been rather generally accepted, there are some exceptions. For example, 2,2'-dimethyl-β-carotene is effective in the rat to the extent of one half of the β-carotene potency [EUGSTER, TRIVEDI and KARRER (*110*)].

The mechanism of this in vivo degradation will be discussed on p. 139. Whereas earlier investigations into the relative strengths of various provitamins A had been restricted to the behavior of all-*trans* carotenoids, newer assays have demonstrated the profound dependence of the bio-potency also on the spatial configuration, hence on the morphological features of the molecule. Survey articles, (*495, 496, 223, 88*). The first observations were made in 1936–1937 by GILLAM et al. (*130, 131*) who found that their "pseudo-α-carotene" (mainly neo-β-carotene B, p. 8) had a provitamin A potency "of the same order" as that of all-*trans*-β-carotene; furthermore, "neo-α-carotene" displayed about 70% of the all-*trans*-β-carotene effect. A few years later KEMMERER and FRAPS (*243*) reported that "pseudo-α-carotene" was half as potent as β-carotene [cf. also BEADLE and ZSCHEILE (*31*)].

Under the guidance of the late H. J. DEUEL, Jr. our group conducted an extended series of pertinent bioassays (1944–1952) in which either crystalline or chromatographically homogeneous samples were used (*89–97, 506, 523, 145*). *Table 18*, p. 119, contains a summary of our results and some obtained by other research groups.

In the course of these and other studies the following fundamental facts have emerged: (a) If an all-*trans* carotenoid shows provitamin A activity, then its *cis* forms are also active. (b) All-*trans* → *cis* re-

Table 18. Relative Potencies of Some Stereoisomeric Provitamins A in the Rat.

Stereoisomeric set	Configuration	Biopotency		References
		in % of that of the all-*trans* form of the set	in % of that of all-*trans*-β-carotene	
β-Carotene	all-*trans*...............	100	100	—
	central-mono*cis* (synth.)	50–30*	50–30*	(*506*)
	peripheral mono*cis*			
	(neo U).............	38	38**	(*95*)
	di*cis* (neo B).........	53	53	(*93*)
	11,11′-di*cis* (hindered)			
	(synth.).............	30	30	(*195*)
	"*cis*C"(hindered) (synth.)	20–25	20–25	(*127*)
β-Carotene	all-*trans* (synth.).......	100	20	(*92*)
homolog C₄₂H₅₈	middle-di*cis* (synth.) ...	100	20	(*92*)
α-Carotene	all-*trans*...............	100	53	(*97*)
	peripheral mono*cis*			
	(neo U).............	25	13	(*97*)
	di*cis* (neo B).........	30	16	(*93*)
γ-Carotene	all-*trans****...........	100	43	(*94, 523, 495*)
	group of *cis* forms.....	60	16	(*91*)
	neo P	70	19	(*91*)
	pro-γ-carotene........	100	43	(*94, 523*)
Cryptoxanthin	all-*trans*...............	100	57	(*146*)
	peripheral mono*cis*			
	(neo U).............	45	27	(*90*)
	mono*cis* (neo A).......	71	42	(*96*)

* In low doses, 50%. Otherwise, the experimental errors do not seem to exceed 10% of the values.

** BAHL et al. (*19*) found 18.5%.

*** Ex pro-γ-carotene, by iodine catalysis. All-*trans*-γ-carotene preparations ex tomato paste or *Pyracantha* berries showed only 26–27% of the β-carotene potency instead of 43%. For possible reasons see p. 75.

arrangements decrease the biopotency (with rare exceptions). (c) There is, in principle, no difference between the biological behavior of unhindered and hindered *cis* carotenoids. (d) The order of magnitude of the influence of stereoisomerization on the biopotency is the same as that of reasonably chosen structural conversions.

To illustrate point (d) let us start from all-*trans*-β-carotene and convert it into all-*trans*-α-carotene by moving a terminal double bond out of conjugation; the biopotency falls from 100% to 53%. This biological weakening is, however, surpassed when, without altering the conjugated β-carotene system, a peripheral double bond is rotated from the *trans* into the *cis* configuration; the resulting neo-β-carotene U shows only 38% of the initial activity.

The conspicuous variation of the individual provitamin A potencies within a stereoisomeric set is evidently the result of numerous partial bioprocesses, whose outcome is the promotion of growth that can be measured in A-depleted young rats. Very little is known about the complicated mechanism of these processes. Contributing factors are, the extent of absorption, stepwise cleavage, destruction and excretion of the provitamin on one hand, and the rates of complexing with, and release from, enzyme systems on the other. The oxidative destruction of provitamins is also influenced by the presence or absence of antioxidants such as tocopherol.

Besides the mentioned growth tests, several other experimental approaches are useful in this field. Thus, according to JOHNSON and BAUMANN (*216*) different doses of stereoisomers are required in order to accumulate equal amounts of vitamin A in the liver and kidneys of the rat.

Such biologically equivalent daily doses are, 20 μg. of all-*trans*-β-carotene, 40 μg. of neo-β-carotene B, and 60 μg. of neo-β-carotene U. The ratios of these A-accumulating effects are, all-*trans* : neo B : neo U = 100 : 48 : 33, in remarkably good agreement with the ratios of the growth-promoting effects established in our laboratory, viz. 100 : 53 : 38.

As mentioned on p. 57, *cis-trans* isomeric carotenoids may undergo bio-stereoisomerization of which a few examples follow.

(a) KEMMERER and FRAPS (*245*, *244*) observed that, upon the ingestion of neo-β-carotene U, half of the carotene derived from rat feces was present in the all-*trans* form. Further, after a Wesson oil solution of neo U had remained in the digestive tract for several hours, the recovered pigment contained, besides unchanged neo U, substantial amounts of the neo B and all-*trans* isomers (see also PAUL et al. *346*). (b) Feeding all-*trans*-lycopene to carotenoid-depleted chickens resulted in the presence of some neo forms in the feces (*89*). (c) DEUEL et al. (*89*) administered pro-γ-carotene and prolycopene to chicks, whereupon the feces, gut washings and livers were investigated, with the result that 50–70% of the recovered pigment showed altered configuration. Interestingly, this bio-stereoisomerization did not lead directly from the starting material to the all-*trans* compound. Indeed, four poly*cis* lycopenes appeared, one of which possessed shorter wavelength maxima and probably contained more *cis* double bonds than prolycopene (cf. p. 71).

When trying to interpret the observations reported above, it would be tempting to assume that a *cis* provitamin A must first rearrange to its all-*trans* form in the body before it may become a proper substrate for the enzymatic cleavage. Such oversimplification would, however, contradict the experimental evidence. We rather propose that the biological dividing line does not run between active *trans* and inactive *cis* compounds but between such isomers whose molecular shape fits into the enzyme system and those that do not satisfy this postulate. Characteristically, all-*trans*-γ-carotene and pro-γ-carotene show equal activities in the rat; and in the chick this *cis* form is even more effective (by 25%) than the all-*trans* isomer (*145*).

XII. Vitamins A and Retinenes.

For more detailed information on this field the following survey articles and monographs should be consulted: AMES (9), DARTNALL (86), KODICEK (248), LOWE and MORTON (289), MOORE (321), MORTON and GOODWIN (325), MORTON and PITT (327), WALD (457, 458, 460, 461), WALD and HUBBARD (473).

1. Structural Relationships.

As is well known, the vitamins A and retinenes are lower isoprenologs of the monocyclic carotenoid pigments; their molecules contain 20 carbon atoms instead of 40. As a special feature they carry functional groups at the end of the aliphatic chain. The structures of the four most important compounds to be discussed appear in the formulas (LI)–(LIV).

The vitamin A (A_1) structure was first proposed, with some reservation, by HEILBRON et al. (156, 158) as well as by KARRER, MORF and SCHÖPP (233) in 1931–32 and was conclusively proved by KARRER et al. (234) in 1933. MORTON, SALAH and STUBBS (328) have demonstrated that the vitamin A_2 molecule contains one more conjugated ring double bond than does A_1. That indeed vitamin A_2 is 3,4-dehydro-vitamin A_1 was shown via synthesis by FARRAR, HAMLET, HENBEST and JONES (115) in 1952; see also (159).

(LI.) Vitamin A_1 (vitamin A, vitamin A alcohol, axerophthol).

(LII.) Retinene₁ (retinene, retinal, vitamin A aldehyde).

(LIII.) Vitamin A$_2$ (3,4-dehydro-vitamin A$_1$, dehydro-retinol).

(LIV.) Retinene$_2$ (3,4-dehydro-retinene$_1$).

The same *numbering system* as in the carotenoid series is used here; another system, with $C_{(1)}$ at the end of the main aliphatic chain, is still preferred by some authors.

In 1934 WALD (*449*) gave the first formulation of the "visual cycle" (p. 143), according to which the visual purple is a complex of the protein "opsin" and the polyene "retinene". Soon afterward, MORTON made the important discovery that retinene is indentical with vitamin A aldehyde [MORTON (*322*); MORTON and GOODWIN (*324*); BALL, GOODWIN and MORTON (*20*)]. For the retinene$_2$ structure see (*328, 115*).

The interconversion, vitamin A $\rightleftarrows$ retinene, can be effected by various agents. Under mild conditions, applying MnO_2 or KBH_4, for example, quantitative yields are obtained without the risk of configurational changes.

Literature: (*20, 45, 53, 167, 180, 231, 324, 328, 390, 455, 472*). For the reversible enzymatic reduction of retinene to vitamin A see e. g. BLISS (*36*) and p. 143.

Closely related to vitamin A are the synthetic parent hydrocarbon axerophthene (LV) (*225*) and the *retro* compound anhydrovitamin A (LVI). The latter is easily obtainable by dehydration of vitamin A at room temperature with HCl in alcohol or chloroform or with *p*-toluenesulfonic acid in benzene (*101, 402, 401, 352*). The *retro* structure of anhydro-vitamin A changes to a *normal* structure on hydrolysis of its BF_3-complex; from the resulting hydroxy-axerophthene the anhydrovitamin A can be recovered by dehydration with HCl in chloroform (*352*).

(LV.) Axerophthene.

(LVI.) Anhydrovitamin A.

2. Number and Character of *cis* Vitamins A and *cis* Retinenes.

The *cis-trans* isomeric vitamins A and retinenes constitute stereo-chemically equivalent sets and can be best discussed jointly. Their exploration has been facilitated by the happy circumstance that, as already mentioned, the interconversions, *cis* vitamin A $\rightleftarrows$ *cis* retinene, can be achieved without configurational changes. Thus, some *cis* vitamins A have been prepared by reducing the corresponding *cis* aldehydes, while in other instances the opposite route was chosen.

As in the closely related carotenoid field, individual *cis* forms are available by isolation from natural sources, by total synthesis, and by in vitro or in vivo stereoisomerization.

In stereoisomeric vitamin A or retinene equilibria only a few "preferred" isomers appear (cf. p. 13); the number of these C_{20} isomers constitutes a higher fraction of all possible *cis* forms than in the C_{40} series. Under certain conditions a sterically hindered configuration may be a preferred one.

For theoretical contributions to pertinent problems see PULL-MAN et al. (*376, 375, 32*); cf. also p. 12.

The "rather tangled story" of the discovery, identification and denomination of the presently known *cis* vitamins A and retinenes has been told by MORTON and PITT (*327*) and will not be repeated here.

Table 19.   Calculated Number of Stereoisomeric Vitamins A and Retinenes.

Type	Sterically unhindered forms	Sterically hindered forms	Total
All-*trans*.	1	0	1
Mono*cis* .	2	2	4
Di*cis*....	1	5	6
Tri*cis* ...	0	4	4
All-*cis*...	0	1	1
Total ...	4	12	16

The configurations of the four theoretically possible sterically unhindered and twelve hindered vitamins A (and retinenes) *(Table 19)* appear in *Fig. 52* (p. 12). At present, the following six forms are known in both sets, including all four possible unhindered forms (tri- and tetra-*cis* isomers are unknown):

all-*trans*   11-*cis* (neo b) (hindered)
9-*cis* (iso a)   11,13-di*cis* (neo c) (hindered)
13-*cis* (neo a)
9,13-di*cis* (iso b)

Melting Points. The simple rule, established in the carotenoid class that the highest melting member of a stereoisomeric set is the all-*trans* compound is invalid in the vitamin A, vitamin A_2 and retinene sets *(Table 20)*. Thus, 9-*cis*-vitamin A melts 18° higher and 13-*cis*-retinene melts 13–14° higher than the corresponding all-*trans* form; 11,13-di*cis*-

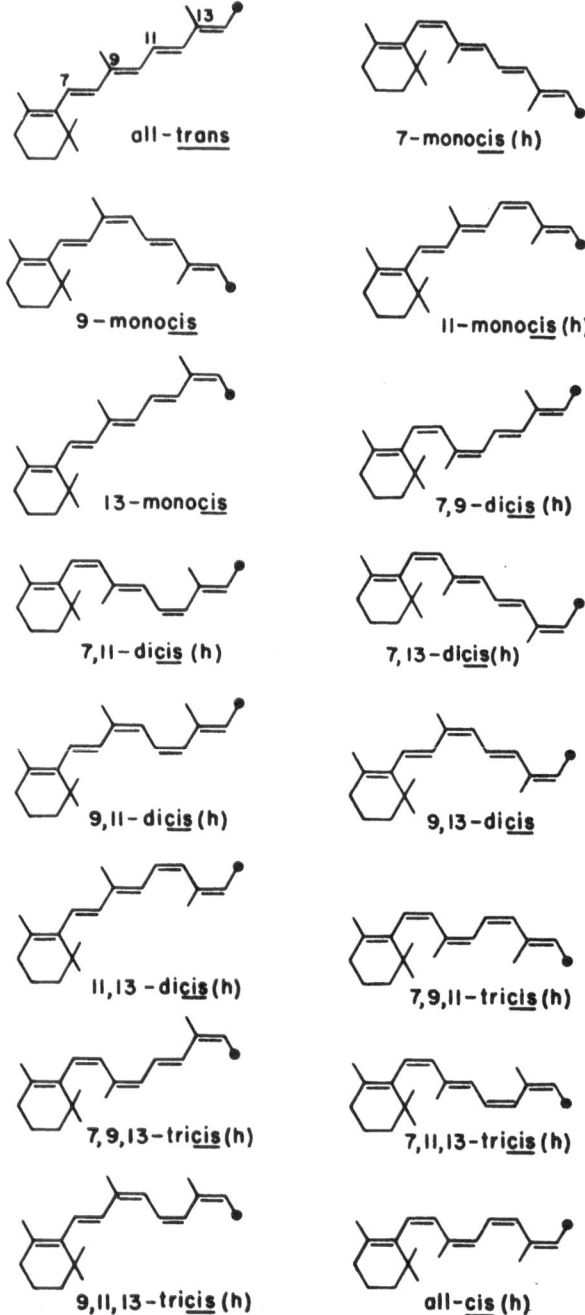

Fig. 52. **Models** of the Sixteen Stereoisomeric Vitamins A and Retinenes. (The dots stand for CH₂OH or CH=O; h indicates a sterically hindered form.)

vitamin A and 11,13-di*cis*-vitamin A_2 represent the highest-melting known members of their respective sets (*209*). All-*trans*- and 9,13-di*cis*-retinenes were obtained in dimorphic forms.

Table 20. Melting Points of Some Vitamin A and Retinene Isomers.

Configuration	Vitamin A set	Retinene set	References
All-*trans*.........	64°	57°; 65°	(*391, 390, 20*)
9-*cis*	81.5–82.5°	64.5°	(*391, 178*)
13-*cis*	58–60°	75°	(*391, 389*)
9,13-di*cis*	58–59°	49°; 85°	(*390*)
11-*cis*	(oil)	63.5–64.5°	(*335, 336, 98*)
11,13-di*cis*	86–88°		(*209*)

Configuration	Vitamin A_2 set	Retinene$_2$ set	References
All-*trans*.........	63–65°	77–78°	(*209*)
9-*cis*	77–79°	54–56°	(*209*)
11-*cis*	(oil)	(oil)	(*209*)
13-*cis*	73–74°		(*209*)
9,13-di*cis*	(oil)		(*209*)
11,13-di*cis*	91–93°	(oil)	(*209*)

Formation and Stability of Hindered Isomers. As pointed out on p. 16, a hindered carotenoid once formed, e. g. by stereospecific half-reduction of a triple bond, shows considerable stability, but in general cannot be prepared by direct rearrangement. This restriction does not hold either in the vitamin A and retinene or in the diphenylpolyene sets (cf. p. 149). Thus, the sterically hindered 11-*cis*-vitamin A and 11-*cis*-retinene can be prepared by rearranging the all-*trans* form (p. 131).

11-*cis*-Retinene was found to be different from all four unhindered isomers synthesized by ROBESON et al. (*391, 390*). Hence, WALD, BROWN, HUBBARD and OROSHNIK (*467, 336*) concluded that it must be a hindered form [cf. also BROWN and WALD (*45*)]. The important problem of its configuration has been solved by OROSHNIK (*335*) who identified WALD's retinene with his totally-synthetic 11-*cis* compound. 11-*cis*-Retinene is the first sterically hindered *cis* polyene found in nature (except perhaps for some poly*cis* carotenoids).

For the synthesis and stability of various hindered *cis* forms of other isoprenic polyenes see OROSHNIK et al. (*337–340*).

3. *Cis-trans* Isomerism and Spectra.

Ultraviolet Spectra. The main band of all-*trans*-vitamin A shows some much discussed special features, viz. the relatively low extinction values and the very moderate fine structure which is absent in the retinene set (*167*) (*Figs. 53–55*, pp. 126, 127). These features are demonstrated by a comparison with the spectrum of the parent hydrocarbon axerophthene (p. 122); the latter shows considerable fine structure and "carotenoid"-like maxima (at *331, 346, 367* mμ, in hexane) whose location approximates the maxima of phytofluene, an aliphatic conjugated pentaene (331, *348*, 367 mμ) (XLII, p. 102). Similarly "abnormal" are the spectra of all-*trans*-vitamin A$_2$ *(Fig. 53)* and the corresponding retinene$_2$. Partial explanation has been offered by OROSHNIK et al. *(337, 340, 339)*, FARRAR et al. *(115)*, and DALE *(79)*. It has been proposed that the character of the fundamental band, indicating a moderately degraded spectrum, is caused by the spatial conflict between the *gem.*

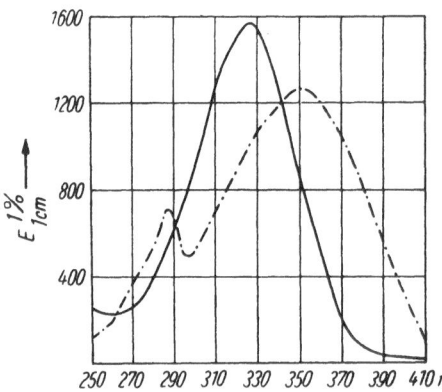

dimethyl group and the 8-hydrogen atom—a feature common to other β-ionylidene and dehydro-β-ionylidene compounds (p. 17). Conversely, the curves of the corresponding *retro* compounds as well as the acyclic C$_{40}$-carotenoids of the phytofluene-lycopene type do possess sharply accentuated fine structure in the main band. So far as this writer can see, the reported spectroscopic difference between vitamin A and axerophthene has remained unexplained.

Fig. 53. Extinction curves in petroleum ether:
————, vitamin A$_1$ acetate; and ·—·—·,
vitamin A$_2$ acetate; according to ISLER *(194)*.
[From: Angew. Chem. 68, 547 (1956).]

Turning to *cis* vitamins A and *cis* retinenes *(Table 21)*, we note that the spectral relationship between the all-*trans* compound and some *cis* isomers is also "abnormal" in the light of the situation found in C$_{40}$-carotenoid sets. Thus, λ_{max} of 13-*cis*-vitamin A is located at a *longer* wavelength (by 3 mμ) than that of the all-*trans* form. Although some

Table 21. Position and Height of λ_{max} in the Vitamin A and Retinene Sets

(in alcohol; u = unhindered, h = hindered *cis* form).

Configuration	Vitamin A set				Retinene set			
	λ_{max} (mμ)	λ_{max} shift (mμ)	$\varepsilon \times 10^{-4}$	References	λ_{max} (mμ)	λ_{max} shift (mμ)	$\varepsilon \times 10^{-4}$	References
All-*trans* .... u	325	0	5.28	*(391)*	381	0	4.34	*(390)*
9-*cis* (iso a) .... u	323	2	4.23	*(391)*	373	8	3.61	*(390)*
13-*cis* (neo a) ...... u	328	(+)3!	4.83	*(391)*	375	6	3.56	*(390)*
9,13-di*cis* (iso b).... u	324	1	3.95	*(391)*	368	13	3.24	*(390)*
11-*cis* (neo b) ...... h	319	6	3.49	*(45)*	376.5	4.5	2.49	*(45)*
11,13-di*cis* (neo c) .. h	312	13	2.62	*(467)*	373	8	1.99	*(467)*

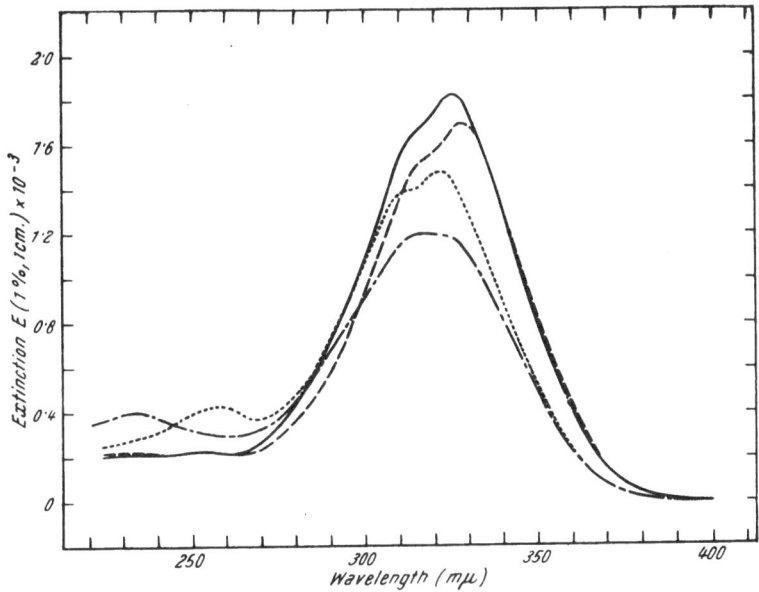

Fig. 54. Extinction curves of some stereoisomeric vitamins A, in hexane: ————, all-*trans*; — — —, 13-*cis*; · —·—· 11-*cis*; and ·····, 9-*cis*; according to HUBBARD (*167*). [From: J. Amer. Chem. Soc. 78, 4662 (1956).]

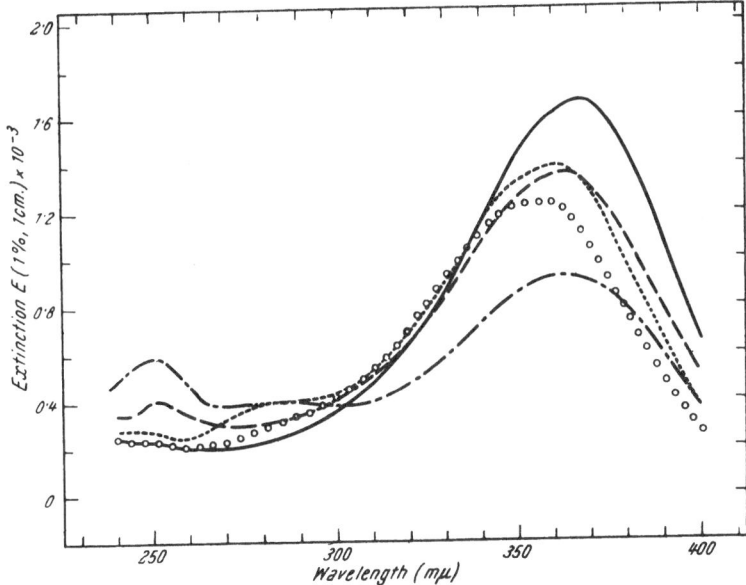

Fig. 55. Extinction curves of some stereoisomeric retinenes in hexane: ————, all-*trans*; — — —, 13-*cis*; · —·—·, 11-cis; ·····, 9-*cis*; and ooooo, 9,13-di*cis*; according to HUBBARD (*167*). [From: J. Amer. Chem. Soc. 78, 4662 (1956).]

other isoprenic *cis* polyenes with comparable chromophores show a similar behavior, it should be stressed that 13-*cis*-retinene behaves "normally" [λ_{max} shift, 6 mμ towards *shorter* wavelengths (*167*)]. Furthermore, the unhindered 9,13-di*cis*-retinene shows a λ_{max} shift of 13 mμ but the hindered 11,13-di*cis* form a much lower value, i. e. 8 mμ. It is also surprising that the λ_{max} shift for 9,13-di*cis*-vitamin A is negligible (1 mμ).

As in C$_{40}$-carotenoid sets, the extinction at λ_{max} of the all-*trans* compound surpasses that of any of the *cis* forms and the values for the

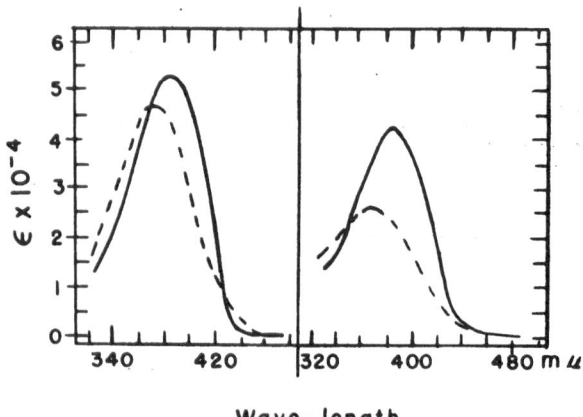

Wave-length

Fig. 56. Influence of extreme low temperatures on the molecular extinction curves of all-*trans*-retinene (left) and the hindered isomer 11-*cis*-retinene (right) in ether-isopentane-alcohol: — — —, at room temperature; and ————, at — 185° to — 193°; according to LOEB, BROWN and WALD (*288*). [From: Nature 184, 614 (1959).]

sterically hindered *cis* isomers are the lowest. Thus, as shown in *Figs. 54* and *55*, the main bands of the hindered 11-*cis* isomers are considerably flattened and have a rather symmetrical shape. Compared with the spectrum of any known unhindered *cis* vitamin A or *cis* retinene these bands are degraded (*167*).

The *cis*-peak effect is observable, as expected, in both sets (*167*). In hexane solution, 9-*cis*-vitamin A shows the peak at 259 mμ, i. e. at about 63 mμ distance from λ_{max} of the all-*trans* form (cf. p. 35); and 9-*cis*-retinene has a peak at 282.5 mμ (distance ~ 80 mμ). The two hindered 11-*cis* isomers do not show distinct peaks, but a moderate elevation in the *cis*-peak region is evident. Catalysis by iodine either lowers or heightens the extinction at *cis*-peak to attain the value observed at quasi-equilibrium.

Retinene Oxime. The fundamental bands of the two unhindered isomers, 9-*cis* and 13-*cis*, display slight fine structure, and the 9-*cis* curve shows a marked *cis*-peak (*167*).

UV Spectra at Extremely Low Temperatures. The pertinent observations made for carotenoids and the explanation given by WALD et al. (*220, 288, 459*) (p. 44) are also valid in the retinene and vitamin A sets. By cooling an all-*trans*-retinene solution to — 185°, the main maximum is displaced by 14 mμ to longer wavelengths and the extinction value is increased by 10%; sterically unhindered *cis* retinenes behave in a like manner. However, in the case of the hindered 11-*cis* form, although the λ_max shift has a similar value (16 mμ), the cooling increases the maximum extinction by as much as 62% *(Fig. 56)*. Furthermore, while the extinguished areas of all-*trans* and unhindered *cis* compounds remain essentially unaffected by the temperature of liquid nitrogen, those of sterically hindered forms, such as 11-*cis*-retinene or 11-*cis*-vitamin A, increase by 40%. In brief, the spectral curves of sterically hindered vitamins A and retinenes undergo "abnormally" profound changes on cooling *(Fig. 57)*, probably because of a partial release of steric hindrance. No considerable fine structure appears.

Pertinent studies of stereoisomeric vitamins A are made difficult by the relatively short wavelength position of the extinguished area and still more by the intense fluorescence at the temperature of liquid nitrogen. Nevertheless, as *Fig. 58* shows, the behavior

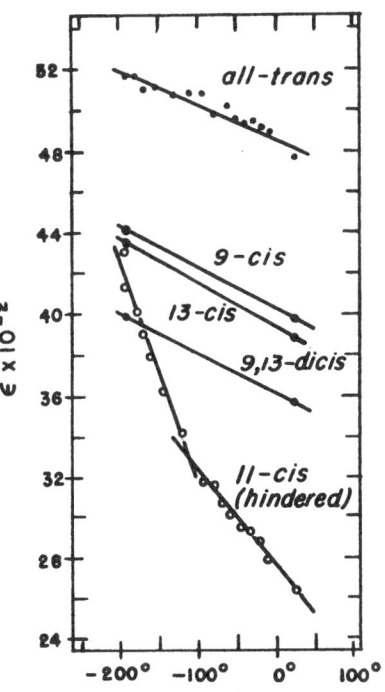

Fig. 57. Increase of the extinction values at λ_max of stereoisomeric retinenes on cooling to the temperature of liquid nitrogen. The rise in the extinction of the all-*trans* and the unhindered *cis* compounds averages 11%, but that of the hindered 11-*cis* isomer is 62%; according to LOEB, BROWN and WALD (*288*). [From: Nature 184, 614 (1959).]

of vitamin A fits well into the general pattern. The same statement is valid for all-*trans*-vitamin A_2 whose weak fluorescence interferes little with spectral measurements.

Infrared Spectra (cf. pp. 79, 81). Five steric forms of retinene whose molecules contain two unmethylated and two methylated C=C double bonds, have been investigated by ROBESON et al. (*390*). The reported data can be only partially correlated with results obtained in the carotenoid sets. The band characteristic of unmethylated *cis* double bonds (12.84–12.95 μ) does not appear in either of the retinene curves, but the expected 10.35 μ band (unmethylated *trans*) is present with high intensity in each of them. A split of this band (*trans-cis* diene, 10.35, 10.47 μ) was observed as expected for the 13-*cis* isomer. No clear statement can be made concerning the 7.25 μ band (methylated *cis*) that should occur in the spectra of the 9-*cis*, 13-*cis* and 9,13-di*cis* forms.

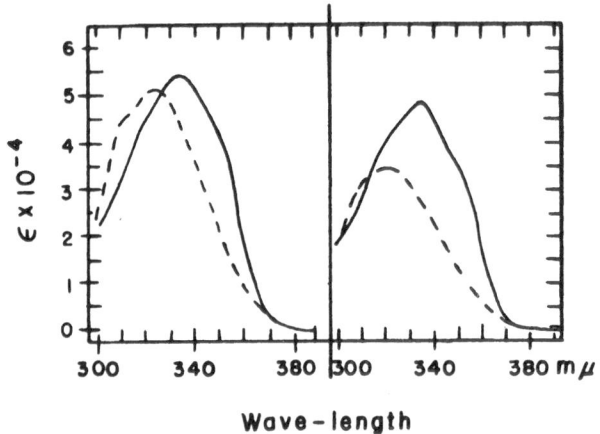

Wave-length

Fig. 58. Influence of extreme low temperatures on the molecular extinction curves of all-*trans*-vitamin A (left) and of the hindered isomer 11-*cis*-vitamin A (right), in ether-isopentane-alcohol: — — —, at room temperature; and——, at — 184° to — 192°; according to LOEB, BROWN and WALD (*288*). [From: Nature 184, 614 (1959).]

The intensity but not the location of the carbonyl stretching band (5.99–6.01 μ) is affected by the configuration.

For the "*cis*" ring double bond in vitamin A_2 cf. FARRAR et al. (*115*).

In a related field OROSHNIK and MEBANE (*339*) have presented extensive IR data referring to aliphatic and *retro* di- to penta-enes. In all instances *trans* configurations were easily confirmed by a strong band at ∼ 10.34 μ but the location of *cis* bands was subject to considerable variation within the 13–15 μ region. No effect of steric hindrance on these IR spectra could be observed.

4. Preparation of *cis* Vitamins A and *cis* Retinenes by Direct Rearrangement.

Although the easy interconversion, all-*trans*-vitamin A ⇄ 13-*cis*-vitamin A (neo a), indicates that thermal methods are effective in these sets, photochemical isomerization methods are mostly used at the present time.

a. Thermal Methods of Stereoisomerization.

In aqueous media (in the presence of 1% digitonin as detergent) heating for 3 hr. at 70° left all-*trans*- and 9-*cis*-retinenes unchanged; the 13-*cis* and 9,13-di*cis* forms rearranged only slightly but 11-*cis*-retinene was converted into the all-*trans* compound almost quantitatively. The relative configurational lability of 11-*cis*-retinene surpasses that of its stereoisomers and of 11-*cis*-vitamin A even at — 15°, in darkness [WALD et al. (*467*)]. Characteristically, when a solution of 11,13-di*cis*-retinene was stored at — 15°, crystals of 13-*cis*-retinene were deposited [OROSHNIK et al. (*336*)].

The much preferred 13-*cis* isomer is also observed in end products of some chemical conversions (*179*).

For stereoisomerization during storage of aqueous multivitamin drop preparations, cf. AMES et al. (*14*, *279*).

b. Photo-stereoisomerization in the Absence of Catalysts.

A number of pertinent data have been reported by HUBBARD (*167*), BROWN and WALD (*45*), WALD et al. (*467*), and OROSHNIK et al. (*336*).

By illuminating hexane solutions, each known *cis* vitamin A or retinene can be converted into the corresponding all-*trans* form almost quantitatively, but in homeopolar solvents such as alcohol the isomerizate contains a much higher proportion of *cis* isomers. In bright sunshine all-*trans*-retinene rearranges partially to give the 11-*cis* and 13-*cis* forms; at extreme dilution the yield of the hindered 11-*cis* isomer may reach 20–25%.

Example. A solution of 10 g. of all-*trans*-retinene in 1.5 l. of abs. alcohol was exposed to bright sunshine for several hours, with stirring, and was then protected from light. By evaporation, fractional crystallization and seeding the 13-*cis* and 11-*cis* forms were obtained; yield, up to 2 g. of 11-*cis*-retinene (*45*).

The photochemical rearrangement of retinene is not prevented either by extremely low temperatures or by the use of rigid solvents: as at room temperature, quasi-equilibria are formed in which the all-*trans* isomer predominates (*220*). In contrast to retinene contained in a glass phase, no photoisomerization has yet been observed in crystals.

Overexposure results in destruction of retinene; filters transmitting only $\lambda > 411$ mμ can be used with advantage (*173*).

c. Stereoisomerization by Iodine Catalysis.

While the various C_{40}-carotenoids show a fairly uniform photochemical behavior in the presence of the catalyst, shorter polyenes have given widely divergent results: some of their *cis* forms rearrange even in darkness but others are iodine-resistant even in light [OROSHNIK and MEBANE (*339*)]. In the vitamin A and retinene sets the rates of stereoisomerization depend not only on the configuration but also on the nature of the functional group, in the sense that any known *cis* retinene rearranges faster than the corresponding *cis* vitamin A (or retinene oxime). This behavior of the aldehydes may be explained by the presence of resonance hybrids such as $\ldots \overset{+}{C}H{-}O^-$, etc. [HUBBARD (167)].

Table 22. Relative Rates of Iodine-catalyzed Photo-isomerization of Some Stereoisomeric Vitamins A and Retinenes, according to HUBBARD (167)

(hexane solutions, white light; u = unhindered and h = hindered forms).

Stereoisomer	Vitamin A set Half-completion time (min.)	Retinene set Half-completion time (min.)
All-*trans* .... u	8	1.7
13-*cis* (neo a) ... u	8	1.3
11-*cis* (neo b) ... h	16	0.5
9-*cis* (iso a) ..... u	~ 400	1.5; 175*
9,13-di*cis* (iso b) .. u	—	1.0; 175*

* These data refer, respectively, to the fast and slow phases of the rearrangement.

Some pertinent data appear in *Table 22*.

According to ROBESON et al. (391), the iodine-catalyzed rearrangement of 13-*cis*-vitamin A acetate takes place (slowly) even in darkness, and the composition at equilibrium is very similar to that observed in light, viz. all-*trans* : 13-*cis* = 2 : 1. The same proportion was found for vitamin A anthraquinone-carboxylate.

Among the known members of the vitamin A and retinene sets the hindered 11-*cis* configuration is the most iodine-sensitive one *(Figs. 59, 60)* and rearranges rather rapidly, even in darkness. In contrast, the unhindered 9-*cis*-vitamin A and 9-*cis*-retinene, when catalyzed with iodine in light, display unexplained stability (167).

As observed in diphenylpolyene and other sets, when a polyene contains more than one *cis* double bond, each of these rearranges under the influence of iodine practically at its "own" rate, i. e. as if it were the sole *cis* double bond present (cf. pp. 172, 179). A similar phenomenon was described recently by HUBBARD (167): Upon catalysis of 9,13-di*cis*-retinene, first the iodine-sensitive 13-*cis* double bond isomerized rapidly, followed by the slow equilibration of the iodine-resistant 9-*cis* double bond, as revealed by spectroscopic readings during the process.

For *vitamin A₂* cf. BARNHOLDT (27).

Retinene Oxime. The behavior of the all-*trans*, 9-*cis*, 11-*cis*, and 13-*cis* forms was studied by HUBBARD (167). The iodine-catalyzed isomerization of the hindered 11-*cis* form was found to be a complex process: First, the all-*trans* oxime accumulated as an "intermediate" whose rate of formation was considerably higher than that of its further isomerization.

d. Stereoisomerization by Acid Catalysis.

In the presence of dilute hydrochloric acid, all-*trans*-retinene affords the unhindered isomers 9-*cis* and 9,13-di*cis*. The yield at equilibrium is 18–23%; and the ratio, mono*cis* : di*cis* = 2–4 : 1 (46).

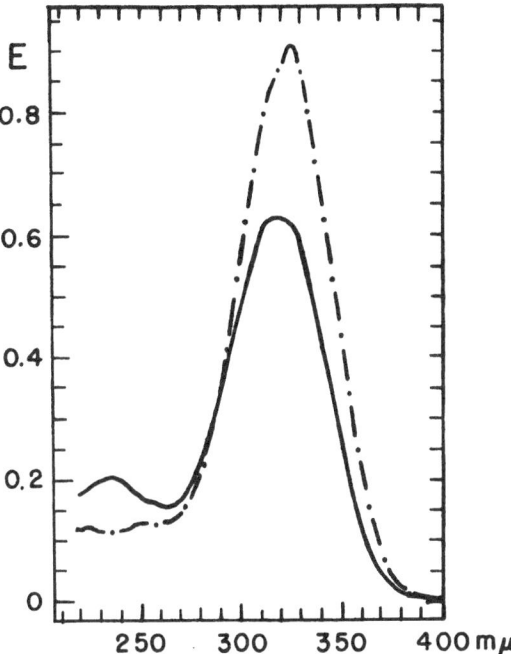

Fig. 59. **Extinction** curve of 11-*cis*-vitamin A, in hexane: ————, fresh solution; and •—•—•, mixture of stereoisomers after catalysis by iodine; according to Brown and Wald (45). [From: J. Biol. Chem. **222**, 865 (1955).]

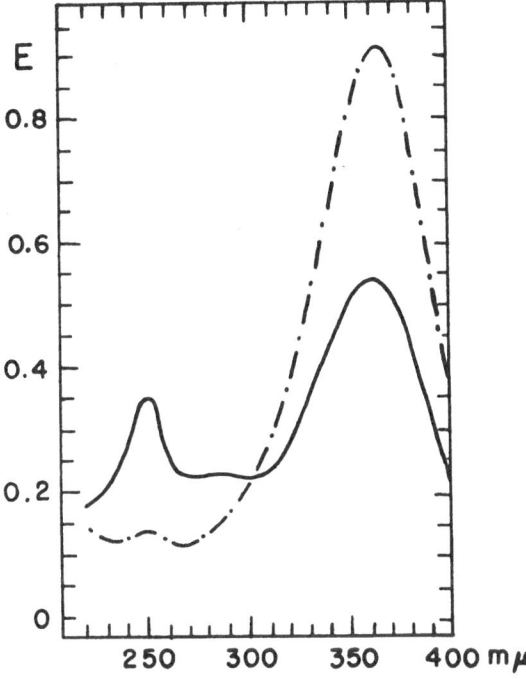

Fig. 60. **Extinction** curve of 11-*cis*-retinene, in hexane: ————, fresh solution; and •—•—•, mixture of stereoisomers **after** catalysis by iodine; according to Brown and Wald (45). [From: J. **Biol.** Chem. **222**, 865 (1955).]

e. Bio-stereoisomerization.

Although no exhaustive studies are available in this field, it is clear that orally given *cis* vitamin A is partially converted into the all-*trans* form and vice versa. AMES et al. (*13*) found, upon feeding to rats either all-*trans*- or 9-*cis*- or 9,13-di*cis*-vitamin A acetate, that a kind of an "equilibrium mixture" appeared in the liver, independent of the spatial form administered. PLACK (*359*) fed 2-mg. doses of 11-*cis*-vitamin A and recovered 1/10 of the dose from the liver; 80–90% of this deposit was in the all-*trans* form. Stereoisomerization has also been observed in liver homogenates (*40, 409, 410*).

For the action of retinene isomerase see p. 144.

5. Preparation of *cis* Isomers by Total Synthesis.

For the synthesis of all-*trans*-vitamin A and -retinene cf. (*157, 200, 199, 208, 210, 315, 317, 16, 124, 179, 30*), and especially the recent survey by ISLER et al. (*209 a*), also discussing syntheses of C^{14}-labeled compounds.

Although the vitamin A chromophore is much shorter than that of the carotenoids, correct stereospecific syntheses of *cis* compounds represent

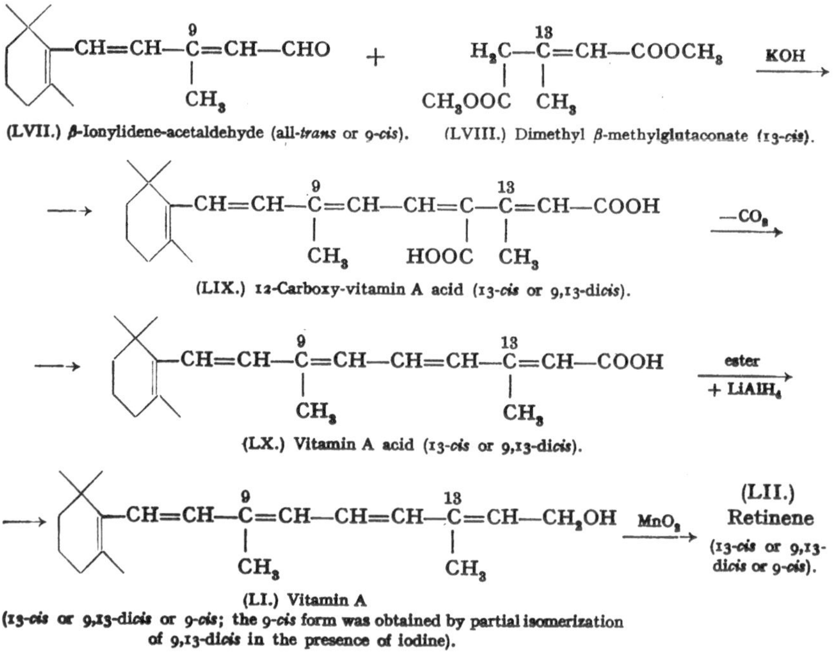

$$-CH=CH-\overset{9}{\underset{CH_3}{C}}=CH-CHO \quad + \quad H_2C-\overset{13}{\underset{CH_3OOC}{\underset{|}{C}}}=CH-COOCH_3 \quad \xrightarrow{KOH}$$

(LVII.) β-Ionylidene-acetaldehyde (all-*trans* or 9-*cis*).   (LVIII.) Dimethyl β-methylglutaconate (13-*cis*).

$$\longrightarrow \quad -CH=CH-\overset{9}{\underset{CH_3}{C}}=CH-CH=\overset{13}{\underset{HOOC}{C}}-\overset{}{\underset{CH_3}{C}}=CH-COOH \quad \xrightarrow{-CO_2}$$

(LIX.) 12-Carboxy-vitamin A acid (13-*cis* or 9,13-di*cis*).

$$\longrightarrow \quad -CH=CH-\overset{9}{\underset{CH_3}{C}}=CH-CH=CH-\overset{13}{\underset{CH_3}{C}}=CH-COOH \quad \xrightarrow[+\ LiAlH_4]{ester}$$

(LX.) Vitamin A acid (13-*cis* or 9,13-di*cis*).

$$\longrightarrow \quad -CH=CH-\overset{9}{\underset{CH_3}{C}}=CH-CH=CH-\overset{13}{\underset{CH_3}{C}}=CH-CH_2OH \quad \xrightarrow{MnO_2}$$

(LI.) Vitamin A

(13-*cis* or 9,13-di*cis* or 9-*cis*; the 9-*cis* form was obtained by partial isomerization of 9,13-di*cis* in the presence of iodine).

(LII.) Retinene (13-*cis* or 9,13-di*cis* or 9-*cis*).

Chart 3. Synthesis of the all-*trans*-, 9-*cis*-, 13-*cis*-, and 9,13-di*cis*-Vitamins A and the Corresponding Retinenes, according to ROBESON et al. (*391*). (The numbering corresponds to that of vitamin A throughout this Chart.)

a rather difficult task. An early synthesis of 9-*cis*-vitamin A was carried out by GRAHAM, VAN DORP and AHRENS (*141*) in 1949. Six years later ROBESON and his group (*391*) synthesized all four sterically unhindered isomers by the route outlined in *Chart 3*, using various spatial forms of the starting materials.

As already mentioned, the hindered 11-*cis*-vitamin A and the corresponding retinene were synthesized by OROSHNIK (*335*) (*Chart 4*, below) who followed a route similar to that recommended for the all-*trans* form by ISLER et al. (*208*).

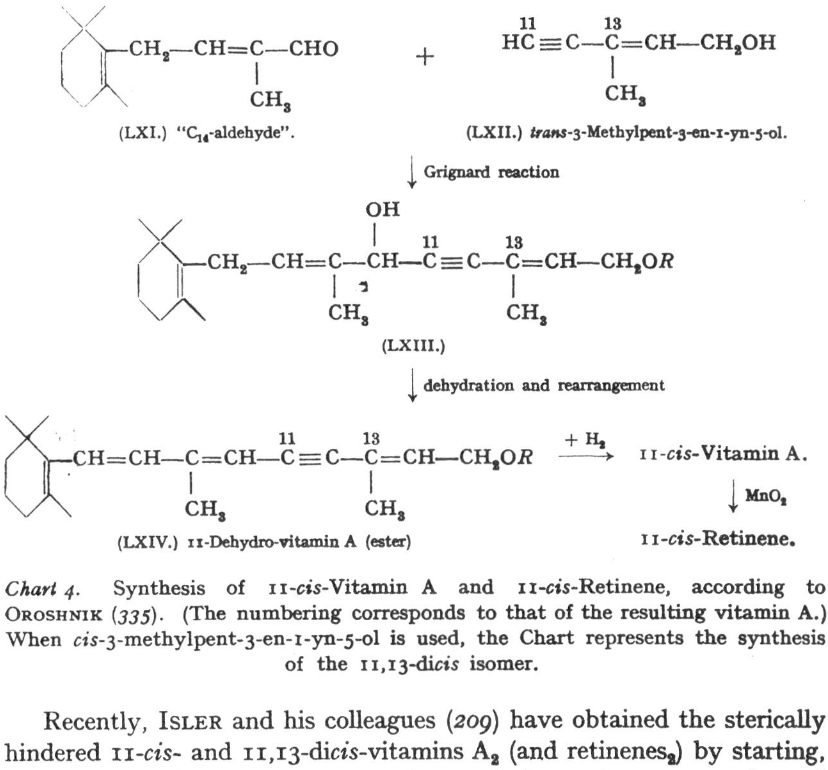

(LXI.) "C_{14}-aldehyde".

(LXII.) *trans*-3-Methylpent-3-en-1-yn-5-ol.

↓ Grignard reaction

(LXIII.)

↓ dehydration and rearrangement

(LXIV.) 11-Dehydro-vitamin A (ester)

$\xrightarrow{+ H_2}$ 11-*cis*-Vitamin A.

↓ MnO_2

11-*cis*-Retinene.

Chart 4. Synthesis of 11-*cis*-Vitamin A and 11-*cis*-Retinene, according to OROSHNIK (*335*). (The numbering corresponds to that of the resulting vitamin A.) When *cis*-3-methylpent-3-en-1-yn-5-ol is used, the Chart represents the synthesis of the 11,13-di*cis* isomer.

Recently, ISLER and his colleagues (*209*) have obtained the sterically hindered 11-*cis*- and 11,13-di*cis*-vitamins A_2 (and retinenes$_2$) by starting, respectively, from *trans*- or *cis*-3-methyl-2-pentene-4-yn-1-ol and dehydro-β-C_{14} aldehyde.

6. Naturally Occurring *cis* Vitamins A and Retinenes.

Vitamins A and retinenes are typical animal products. Nevertheless, trace amounts of retinene are widespread in plants and have been identified in clover, grass, spinach, kale, cress, rose hips, citrus fruits, etc. by means of the extremely sensitive rhodanine reaction [WINTERSTEIN and HEGEDÜS (*481*)]. So far no information is available on the configuration of plant retinene.

The biosynthesis of vitamins A, like that of carotenoids leads preponderantly, but not exclusively, to all-*trans* forms. The first natural *cis* compound was isolated in 1945 by ROBESON and BAXTER (*388, 389*) from shark liver oils. This "neovitamin A" (13-*cis*) also occurs in synthetic concentrates, and the ratio, all-*trans* : *cis*, is 1.5–2 : 1 in both instances. "It appears that 'vitamin A' physiologically speaking must be considered as a mixture of the two geometrical isomers" [CAWLEY et al. (*56*); cf. (*155, 372*)].

In order to isolate natural 13-*cis*-vitamin A, major amounts of the all-*trans* form were eliminated by cooling the ethyl formate solution of a concentrate to — 70°. By chromatographing the petroleum ether solution on sodium aluminum silicate the lower section of a mixed zone was enriched in the *cis* form. The content of the latter section was esterified with phenylazobenzoyl chloride and, after crystallization of the ester and saponification, the liberated *cis* vitamin was crystallized from ethyl formate.

Besides the 13-*cis* compound two other unhindered *cis* vitamins A, 9-*cis* and 9,13-di*cis*, have been encountered in fish liver oils (*46*).

Concerning the sterically hindered 11-*cis*-vitamin A and -retinene in the Vertebrate eye the reader is referred to p. 142.

In Crustacea the vitamin A content, concentrated in the eyes, consists entirely or predominantly of the 11-*cis* isomer. According to WALD and BURG (*470*) the lobster *(Homarus americanus)* carries practically 100% of its vitamin A in this form. In various Eucarida Crustaceae 11-*cis*-vitamin amounts reportedly to 50–90% of the total vitamin A, accompanied in some instances by smaller amounts of the 11,13-di*cis*, 13-*cis* and 9,13-di*cis* forms (*118, 117*; see also *28, 465*). In such stereo-isomeric mixtures all-*trans*-vitamin A participates only to the extent of 10–40%.

It should be mentioned at this point that the occurrence of retinene is not restricted to eyes; in small amounts it has also been observed (in bound form) in egg yolk and in the eggs of some marine teleosts (herring) (*360, 362*). Studies into the stereochemistry of such complexes do not seem to be available.

The presence of retinene in mammal livers had been overlooked for a long time but recently as much as 1-2 mg./kg. was found in cattle liver (*481*). A pertinent configurational study is still lacking. It is not known how far such depots may supply the eye with retinene.

As in the carotenoid class (p. 63), not every recorded occurrence of a *cis* vitamin A or *cis* retinene should be accepted at face value, because of possible rearrangements during isolation and analysis; this is especially true for commercially processed liver oils (*46, 276*). When extracting and purifying carotenoids, the determined all-*trans* values may be found too low, because of *trans* → *cis* isomerization; in the field of vitamins A and retinenes, however, an error in the opposite direction may also occur, especially in the presence of the natural 11-*cis* isomer, when light is not excluded throughout the operations.

With reference to *cis* vitamins A_2 CAMA et al. wrote (*54*):

"It is highly probable that natural vitamin A_2 is a mixture of *cis-trans* isomerides and that the crude retinene$_2$ will also be a mixture ... It may well be that on standing the isomers tend to give an equilibrium mixture and that recrystallization merely accumulates the least soluble isomer." (Cf. also *1*.)

7. Chromatographic Separation of Stereoisomers.

The separation of stereoisomeric vitamins A and the characterization of the individual *cis* forms is more laborious than in carotenoid sets because of greater difficulties in differentiating zones and in crystallization, and because of less specific spectra. In a few instances molecular distillation has been used (*389*). The retinenes can be handled more easily considering the formation of colored zones and greater tendency to crystallize.

The presence of a functional group allows good chromatographic separation of some derivatives such as vitamin A acetates, phenylazo-benzoates or anthraquinone carboxylates as well as retinene oximes, 2,4-dinitrophenylhydrazones, etc. (*391, 167, 20*).

In both the vitamin A and retinene sets the all-*trans* compound shows the strongest adsorption affinity; "neo U" *cis* types are unknown. A hindered *cis* form occupies a lower position on the column than a comparable unhindered one. Of course, any vitamin A_2 isomer will be adsorbed above the sterically corresponding A_1 zone.

Examples. (a) The ether-containing light petroleum solution of an unsaponifiable cod liver oil residue was washed through a dicalcium phosphate column. Under the guidance of the fluorescence of the fractions, 13-*cis*-A_1, all-*trans*-A_1, and all-*trans*-A_2 were collected, in this sequence. The fluorescence of the 13-*cis* form was yellowish-green versus the bluegreen fluorescence of all-*trans*-vitamin A_1; A_2 showed a weaker, brownish fluorescence (*44, 323*). 13-*cis*-Vitamin A_2 was adsorbed between 13-*cis*-A_1 and all-*trans*-A_1 (*27*).

(b) A crude sample of 11-cis-retinene was developed with petroleum ether + 10% benzene on weak alumina. The lowest, orange zone (13-*cis* + 11-*cis*) was eluted with petroleum ether + 1% ethanol. On rechromatographing, a diffuse yellow zone appeared of which the first fraction to run through contained the 11-*cis*-retinene as shown by spectroscopic readings (*45*).

(c) Alumina, partially deactivated with water or HCl has been used repeatedly (cf. e. g. *178, 339, 20, 85, 25*). For sodium silico-aluminate columns see (*391*).

8. Some Methods used in Configurational Assignments.

As in the carotenoid class, the best method is identification of a chromatographically homogeneous isomer with a sample prepared by stereospecific total synthesis. Of course, important conclusions may also be drawn from the physical and chemical behavior of a given isomer.

Spectra. The spectroscopic estimation of the number of *cis* double bonds is less simple here than in carotenoids, considering the "anomalous" λ_{max} shift values mentioned on p. 126. It is easy to differentiate, however, between sterically unhindered and hindered configurations because of the dramatic increase of the extinguished area upon the addition of iodine, in case of steric hindrance. Furthermore, the remarkable, unexplained stability of the spectrum on iodine catalysis differentiates the 9-*cis* configuration from all the others. As pointed out by Oroshnik *(335)*, the absence of the iodine-stable 9-*cis* form in an isomerizate proves the absence of a 9-*cis* double bond in the starting material.

Proton Resonance Spectra have been obtained recently by Kofler and Rubin *(250 a)* for five vitamin A stereoisomers. The same authors have also demonstrated that *X-ray powder diagrams* show marked differences in lattice spacing between the respective steric forms.

Among the *chemical conversions* of vitamin A the well-known Carr-Price reaction cannot be used for the differentiation of stereoisomers because all known steric forms seem to yield the same blue antimony complex *(389, 173, 45)*.

Esterification Rates. The relative rates of enzymatic esterification of vitamin A with higher fatty acids were found as follows: all-*trans*, 100%; 11-*cis*, 97%; 13-*cis*, 60%; and 9-*cis*, 42% *(252)*.

Dehydration. Although the rate of anhydrovitamin A formation (p. 122) is definitely lower for 13-*cis*- than for all-*trans*-vitamin A *(402, 401, 389, 26)*, no pertinent general rules are available at present. In any case the sharp UV maximum of the anhydro product allows the estimation of individual isomers or stereoisomeric mixtures, even in impure solutions.

Maleic Anhydride Adducts. As is widely known, the formation of diene-maleic anhydride adducts is a stereospecific phenomenon *(6, 335)*. Although no important use of this feature has yet been made in the carotenoid class, maleic anhydride did become a convenient tool in the configurational study of vitamins A and retinenes. The interaction of vitamin A and maleic anhydride was first reported by Kawakami *(241)*. Later, Robeson and Baxter *(389)* found that 13-*cis*-vitamin (natural neo a) is much less reactive than the all-*trans* compound. In the adduct, they established the ratio, vitamin A : maleic group = 1 : 1 [Robeson et al. *(391)*].

According to the same authors the spectrum of the adduct (LXV) ex all-*trans*-vitamin A acetate is practically identical with that of β-ionylidene-ethanol acetate (cf. LXVI). This demonstrates that the

(LXV.) Vitamin A-maleic anhydride adduct. (LXVI.) β-Ionylidene-ethanol.

two adjacent terminal double bonds 11 and 13 have participated in the adduct formation; the triene system that is also present in β-ionylidene-ethanol, does not interact with maleic anhydride.

In order to show considerable reactivity both the 11- and 13-double bonds in vitamin A (or retinene) must be *trans* as confirmed by the much reduced reactivity of 13-*cis*-vitamin A. This stereochemical postulate is fulfilled only by four of the sixteen possible members of the vitamin A or retinene sets, viz. the all-*trans*, 7-*cis*, 9-*cis*, and 7,9-di*cis* forms; and by only two of the known six isomers, viz. all-*trans* and 9-*cis* (iso a). The two latter vitamins (and the corresponding retinenes) are termed "fast reacting" and the others "slow reacting" forms.

When, under standard conditions, at room temperature, the maleic adduct formation is interrupted after 30 minutes, the practically unchanged "slow" isomers can be extracted; then the "fast" reacting ones are liberated from their crystalline adducts by means of alcoholic KOH and extracted in turn. Thus, the separation and estimation of the two types becomes possible in artificial mixtures or unsaponifiable residues of liver oils, subject to an error of only a few per cent. The two main fractions can be further resolved chromatographically and/or studied by other methods (*52, 323, 44, 58, 311, 361, 358, 46, 40, 10*).

Opsin Test. As will be seen on p. 143, a test with a degree of specificity otherwise unheard of in the polyene class is available for 11-*cis*-retinene, the only spatial form that combines with the retinal protein opsin to yield the deeply colored rhodopsin. Moreover, in the absence of 11-*cis*-retinene, the 9-*cis* or 9,13-di*cis* isomer can be identified spectroscopically and estimated upon incubation with opsin when the unnatural chromo-proteid isorhodopsin (p. 143) is formed. 9,13-Di*cis*-retinene undergoes slow rearrangement to 9-*cis*-retinene during the test [Hubbard et al. (*173*), Brown et al. (*46*), Brown and Wald (*45*)].

9. Systemic Bioeffects of *cis-trans* Isomeric Vitamins A and Retinenes.

See also the survey articles listed on p. 121, especially (*9, 248,* and *289*).

Biochemical Relationship to Carotenes.

As is well known, the curative effects of provitamins and vitamins A in case of hypovitaminosis are qualitatively identical, because the

C_{40}-carotenes must be converted in the body into C_{20}-vitamins before they are able to act. The mechanism of this cleavage is not entirely clear and can be discussed here only briefly [survey, LOWE and MORTON (289)]. Two well established facts stand out: a) that one mole β-carotene produces the growth effect of one mole and not of two moles of vitamin A; and b) that the cleavage process does not begin, as assumed earlier, at the central double bond which has the least double bond character.

It has been proposed that the attack starts at one of the terminal groups, wherefrom it would proceed by β-oxidation-like steps toward the center and would stop when the retinene stage (C_{20}) is reached. Then the retinene is reduced to vitamin A at high rates [GLOVER and REDFEARN (134), FISHWICK and GLOVER (119), GLOVER, GOODWIN and MORTON (132, 133)]. As this theory requires, the expected intermediates such as β-apo-8'-carotenal (LXVII, see 46 a), β-apo-10'-carotenal, and β-apo-12'-carotenal show strong growth effects in A-depleted young rats (cf. 306 a). Furthermore, apo-10' is converted to apo-12' in vivo [EULER, KARRER and SOLMSSEN (113), GLOVER and REDFEARN (134), FAZAKERLEY and GLOVER (116)]. Recently, β-apo-8'-carotenal was found by WINTERSTEIN et al. (481 a) in the intestinal mucosa of the horse.

In spite of these observations, as was pointed out by GLOVER (131 a) in his comprehensive new review, "the true nature of the conversion process is still far from clear". Thus, the "abnormally" high relative biopotencies of the C_{25}-β-apo-carotenoids remain unexplained (see also 119).

(LXVII.) β-Apo-8'-carotenal (C_{30}).

Bioeffects and Stereoisomerism.

The methods used in the evaluation of the relative potencies of cis-trans isomeric vitamins A and retinenes are the same as in the case of the C_{40}-provitamins A, viz. growth tests, liver storage tests and vaginal-cornification tests (for details see 103). Comparative data obtained respectively in growth tests and liver storage tests have been reported in detail by AMES and his colleagues (11, 12) (Table 23). Some bioeffects have also been correlated with maleic anhydride adduct values (10).

The following general conclusions drawn for carotenes (p. 118) are valid in the vitamin A and retinene sets too: Each known cis isomer is biopotent but acts more weakly than the corresponding all-trans compound; furthermore, there is no qualitative difference between the behavior of sterically hindered and unhindered isomers in the animal body.

That all-*trans*-retinene and all-*trans*-vitamin A show approximately equal potencies points to a near-quantitative bio-reduction of the aldehyde group [WENDLER et al. (*479*)]. No such simple relationship exists, however, between a *cis* retinene and the corresponding *cis* vitamin A (Table 23). Evidently, the situation is complicated here by an intricate pattern of steric rearrangements and other, still unknown processes.

Table 23 shows the following interesting features: a) 11-*cis*-retinene is twice as potent as 11-*cis*-vitamin A acetate. b) While the unhindered 9-*cis*- and the hindered 11-*cis*-vitamins A are equally potent, 9-*cis*-retinene is less than half as active as 11-*cis*-retinene. c) The most active *cis* forms in both sets are the 13-*cis* isomers (terminal *cis*) which may we explained by the easy interconversion, 13-*cis* ⇄ all-*trans*.

Table 23. Relative Vitamin A Potencies of Some Stereoisomeric Vitamin A Acetates and Retinenes in the Rat (Growth Test), according to AMES, SWANSON and HARRIS (*11, 12, 9*).

(All-*trans*-vitamin A acetate = 100%.)

Configuration	Biopotency in A-depleted rats	
	Vitamin A acetate (%)	Retinene (%)
All-*trans*	100	91
9-*cis*	21	19
11-*cis*	23	48
13-*cis*	75	93
9,13-di*cis*	24	17
11,13-di*cis*	15	31

It is reassuring to note that the data listed in Table 23 are in reasonably good agreement with the results obtained by other authors who in part made use of different assay methods. Thus, for 13-*cis*-vitamin A the following data are also available: 81% by the rat-growth method; 72% by the liver storage method (*151*); and 68% by the vaginal-cornification method (*66*).

The response of chicks to the administration of 9-*cis*- or 9,13-di*cis*-vitamin A is very similar to that of rats (*11*). For vitamin A ether see (*317*), and for α-vitamin A ether (*334*).

10. Some Stereochemical Aspects of Visual Processes.

See the surveys listed on p. 121, especially (*86, 321, 325, 327, 457, 458, 460, 461,* and *473*).

The now classical studies carried out by WALD and his school were initiated by the observation (1933) that sizable amounts of vitamin A are present in various eye tissues, especially in retinal rods and cones (*448* to *450*). WALD has discovered that visual purple, termed rhodopsin, is a chromoproteid that contains a specific protein, opsin (in rods, scotopsin and in cones, photopsin), linked to retinene, the prosthetic group. As already mentioned, retinene was identified by MORTON as vitamin A aldehyde (p. 122). The retinene residues are attached to nitrogen atoms of the opsin (326).

In light, rhodopsin dissociates into its two components. The retinene thus liberated is reduced to vitamin A that participates in the subsequent re-formation of rhodopsin. These physiological processes can be duplicated in vitro: when opsin, vitamin A, alcohol dehydrogenase and DNP are mixed, the solution turns purple in darkness but bleaches in light. The only photochemical step in this cycle is the dissociation of rhodopsin [WALD et al. (453, 454, 456, 466, 463); PITT and MORTON (356, 357)].

COLLINS, GREEN and MORTON (68, 69) have achieved 100% regeneration of rhodopsin in vitro.

The fundamental pattern just described is in principle the same in all vertebrates, to which our main discussion is limited.

According to WALD (454) the visual system of fresh-water and some anadromous fishes carries retinene$_2$ and vitamin A$_2$; thus the primary difference from the rhodopsin system is the presence of one more double bond in the retinene$_2$ moiety.

The four visual chromoproteids found in vertebrates are listed in Table 24.

Table 24. Visual Chromoproteids in Vertebrates (473, 469).

Name	λ_{max} (mμ)	Components	
		polyene	protein
Rhodopsin	500	retinene$_1$	rod opsin (scotopsin)
Iodopsin	562	retinene$_1$	cone opsin (photopsin)
Porphyropsin	522	retinene$_2$	rod opsin
Cyanopsin........	620	retinene$_2$	cone opsin

At first, there was no apparent connection between visual systems and the stereochemistry of polyenes. Subsequently, however, it came to light that visual processes are deeply influenced, indeed governed, by the geometrical forms of the vitamin A and retinene that participate in the system. The significance of this discovery becomes evident when Charts 5 and 6 are compared.

The first contact between the two fields was established by HUBBARD and WALD's observation (176–178) that the in vitro re-synthesis of rhodopsin (176) from vitamin A and opsin was successful only when vitamin A ex fish liver oil was used, but it remained unsuccessful when pure crystalline all-trans-vitamin A was the starting material. The same discrepancy was observed when first the two vitamin A preparations were converted by MnO$_2$-oxidation to the corresponding retinenes and these were reacted with opsin [BROWN and WALD (45)].

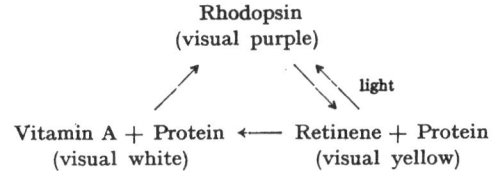

Chart 5. The Rhodopsin Cycle according to WALD (1934–1936) (*451*, *452*, *449*).

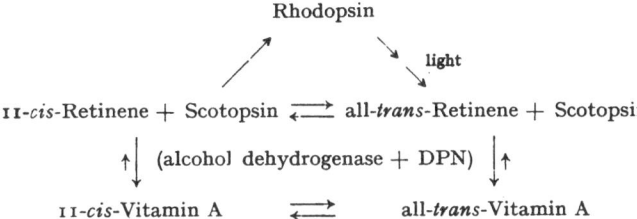

Chart 6. The Rhodopsin Cycle according to WALD and HUBBARD (1960) (*473*). In this whole cycle the only energy-consuming step is the re-isomerization of all-*trans*-retinene to the 11-*cis* form; and the only photochemical process is the stereo-isomerization of the 11-*cis*-retinene moiety of rhodopsin (*461*, *462*).

Illumination of pure all-*trans* preparations resulted in "active" samples, well fit for rhodopsin synthesis, because during this photochemical rearrangement 11-*cis*-vitamin A (or 11-*cis*-retinene) was formed, the latter having a special affinity for opsin. 11-*cis*-Retinene was first isolated by HUBBARD (*165*) who found that it reacts rapidly and completely with opsin.

Further research has shown that those members of the retinene set whose molecules have a straight overall form (all-*trans*, 13-*cis* and 11,13-di*cis*, models on p. 124) are unable to combine with opsin. The 9-*cis* isomer, however, whose molecules are as strongly bent as those of 11-*cis*, couples with rod or cone opsin easily to yield the chromoproteids isorhodopsin or isoiodopsin. The same result obtains with 9,13-di*cis*-retinene, because under the conditions of the incubation it rearranges to the 9-*cis* form. 9-*cis*-Retinene$_2$ yields the pigments isoporphyropsin and isocyanopsin (*468*, *473*). No "iso" pigment has yet been found in nature.

The stereochemical changes that take place during the visual process can be outlined as follows.

A characteristic feature of Chart 6 is that, although 11-*cis*-retinene is required for rhodopsin biosynthesis, the photochemical cleavage of rhodopsin yields exclusively (or perhaps preponderantly) all-*trans*-retinene [HUBBARD et al. (*170*, *174*, *253*)]. As has been revealed by LYTHGOE (*295*) this process includes several steps; he characterized an intermediate as "transient orange". Much later, WALD and

HUBBARD (*473*) have concluded from their experiments that the first phase of the rhodopsin bleaching is a photochemical 11-*cis* → all-*trans* rearrangement to the all-*trans* chromoproteid "lumi-rhodopsin". This step eliminates the specific stabilizing influence of the 11-*cis*-retinene moiety and weakens the polyene-protein link. Still more weakening takes place during the next (thermal) step, to wit the conversion, all-

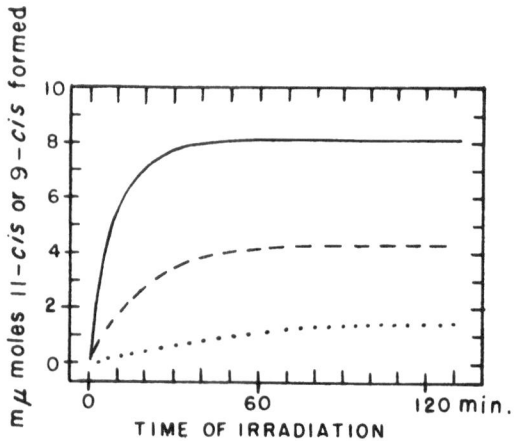

Fig. 61. Influence of retinene isomerase on the photoisomerization of all-*trans*-retinene: top, formation of 11-*cis*-retinene in the presence of isomerase; middle, formation of 11-*cis*-retinene in the absence of isomerase; and bottom, simultaneous formation of 9-*cis*-retinene in the presence or absence of isomerase (no effect); according to HUBBARD (*168*). [From: J. Gen. Physiol. 39, 935 (1956).]

trans-lumi (red-violet) → all-*trans*-meta (orange-red) pigment; the latter proteid releases all-*trans*-retinene (yellow) spontaneously at temperatures above —20° (*471*; cf. also *171*, *172*, *175*).

The sharp dependence of the protein-polyene bond's stability on the retinene configuration is reminiscent of the dependence of the adsorption affinities of stereoisomeric carotenes on the molecular shape.

Although on illumination of a rhodopsin solution the polyene moiety is attacked first, the exposure of rhodopsin to heat attacks the protein moiety and leaves the retinene configuration unchanged [HUBBARD (*169*, *170*); WALD and BROWN (*464*)]:

$$\text{Rhodopsin} \begin{cases} \xrightarrow{\text{light}} \text{all-}trans\text{-retinene} + \text{native opsin} \\ \xrightarrow{\text{heat}} \text{11-}cis\text{-retinene} + \text{denatured opsin} \end{cases}$$

Retinene Isomerase [HUBBARD (*166*, *168*, cf. *473*)]. A fascinating event in the evolution of polyene stereochemistry was the detection in the retina of a photoisomerase, termed "retinene isomerase", that shifts the photochemical quasi-equilibrium, all-*trans*-retinene ⇄ 11-*cis*-retinene, to the right side. The enzyme acts in neutral phosphate buffer, at ∼ pH 6.7, in light; in darkness its effect is very slow.

As indicated on p. 131, irradiation of aqueous all-*trans*-retinene solution (+ digitonin), affords, at moderate rates, a stereoisomeric mixture which contains about 18–25% of the 11-*cis* form (besides other *cis* isomers). In contrast, the presence of isomerase causes, even in dim light, the rapid formation of 35–40% 11-*cis*-retinene as practically the only *cis* form *(Fig. 61)*.

The substrate stereospecificity of retinene isomerase is sharp: neither all-*trans*- nor 13-*cis*-vitamin A is rearranged enzymatically (in the presence of dehydrogenase) into the 11-*cis* form, nor are the 13-*cis*- and 9-*cis*-retinenes. Fig. 61 demonstrates the passivity of 9-*cis*-retinene in contrast to the behavior of 11-*cis*.

Finally, it should be mentioned that in visual processes there exists a parallelism between the respective functions of retinene isomerase and opsin. Since opsin binds 11-*cis*-retinene but releases all-*trans*-retinene, it may be considered as an isomerase too. According to WALD and HUBBARD *(473, 170)* it completes the following in vivo isomerization cycle:

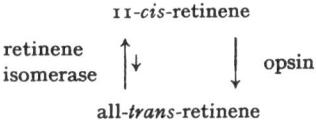

Second Part

Unbranched Arylpolyenes.

Introductory Remarks.

In the following Chapters we will discuss some unbranched polyene systems containing one or two conjugated aromatic terminal groups. The *cis* forms of these compounds can be divided into sterically hindered and unhindered types. While in the carotenoid and vitamin A classes hindrance is caused mainly by the conflict of a hydrogen atom and a methyl side-chain (p. 13), in *cis* α-mono- or α,ω-diaryl-polyenes two hydrogens may overlap. As *Figure 62* shows, when a terminal aliphatic

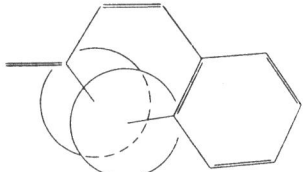

Fig. 62. Overlapping of hydrogen atoms in a peripheral *cis* phenylpolyene. Values used: C=C, 1.33 Å; C—C, 1.46 Å; C$_6$H$_5$—C, 1.44 Å; C$_6$H$_5$—H, 1.08 Å; C—H, 1.09 Å; H radius, 1.20 Å; angles C=C—C and C=C—H, 124° 20' (*514*). [From: J. Amer. Chem. Soc. **64**, 2755 (1942).]

double bond assumes the *cis* configuration, a spatial conflict arises between an *ortho* ring-hydrogen and an H-atom of the aliphatic chain. Thus, a planar or approximately planar configuration becomes impossible; the calculated deviation amounts to 52.5° (*514*).

In the 1-phenylbutadiene, 1,4-diphenylbutadiene and stilbene sets only hindered *cis* forms are possible.

It should be stressed again that the terms "hindered" and "unhindered" *cis* isomers are not synonymous with "preferred" and "non-preferred". Thus, stereoisomeric equilibria may contain marked amounts of hindered *cis* forms, especially in some diphenylpolyene sets. The tendency to form hindered isomers on direct rearrangement is evidently stronger here than in the carotenoid class.

XIII. 1-Monoaryl-polyenes.

$$C_6H_5 \cdot (CH=CH)_nH$$

The stereochemical relationships in the extended class of mono-aryl-polyenes will be illustrated by two very different substances: 1-Phenylbutadiene is a lower molecular weight, volatile liquid; hence,

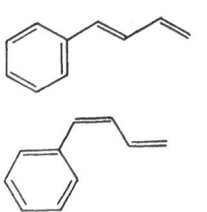

the separation of its spatial forms has been carried out mainly by distillation and by treatment with maleic anhydride (cf. 6); in contrast, the separation of the higher molecular weight, non-volatile polyene, 1-phenyl-undecapentaenal set has required chromatographic methods.

Fig. 63. Models of stereo-isomeric 1-phenylbuta-dienes: top, *trans*; and bottom, *cis*.

1-Phenylbutadiene Set.

The only (hindered) *cis* form of this compound *(Fig. 63)* was isolated by GRUMMIT and CHRISTOPH *(147, 148)* who exposed the *trans* form to sunshine or to an UV-rich artificial light source. Later, WITTIG and SCHÖLLKOPF *(483)* found that about one half of the phenylbutadiene (LXVIII) that originated from the following reaction was in the *cis* configuration and could be isolated as such:

$$(C_6H_5)_3P=CH-CH=CH_2 + C_6H_5 \cdot CHO \rightarrow (C_6H_5)_3PO + C_6H_5 \cdot CH=CH-CH=CH_2$$

(LXVIII.) 1-Phenylbutadiene.

GRUMMIT and CHRISTOPH reported that catalysis by iodine did not markedly isomerize the all-*trans* compound. It is interesting to note that the sterically unhindered methyl analog, *trans*-piperylene, $CH_3-CH=CH-CH=CH_2$, does yield an equilibrium mixture with iodine (14% *cis*) but does not respond to UV irradiation *(121)*.

The isolation of *cis*-phenylbutadiene was a rather laborious procedure. After vacuum distillation, the unchanged portion of the *trans* form was removed with maleic anhydride (the *cis* isomer reacts very slowly), a trace of methylstyrylcarbinol was eliminated by means of methylmagnesium bromide and, after re-distillation, the substance was washed repeatedly through an alumina column.

Some characteristics of the two spatial forms appear in *Table 25.*

Table 25. Comparison of *cis*- and *trans*-1-Phenylbutadienes (*147, 148*).

Configuration	Freezing point	Distillation temp. (11 mm.)	n_D^{25}	λ_{max} in ethanol (mμ)	Reaction with maleic anhydride	Kept at 80° for 15 days
cis	−57°	71°	1.5822	265–266	slow	polymerizes
trans	4.5°	83°	1.6089	280	faster	dimerizes

The strong spectral effect of the *trans* → *cis* rotation is shown in *Fig. 64*; the λ_{max} shift is 14–15 mμ. The IR spectra are clearly different in several regions but a detailed analysis of the curves is not available.

1 - Phenylundecapentaenal-(11) Set.

All-*trans*-phenylundecapentaenal, $C_6H_5 \cdot (CH{=}CH)_5 \cdot CHO$ *(Fig. 65)*, a representative of the vinylogous series $C_6H_5 \cdot (CH{=}CH)_n \cdot CHO$, was first obtained by KUHN and WALLENFELS (*268*). (M. p. 185–186°.) The molecules of this unsymmetrical polyene may assume 32 configurations.

GANSSER and the writer (*122*) have found that when the aldehyde had been submitted to some thermal or photochemical treatments, four new zones (*cis*-I–IV) appeared on the column below the all-*trans* zone. This resolution was successful only on special adsorbents because of

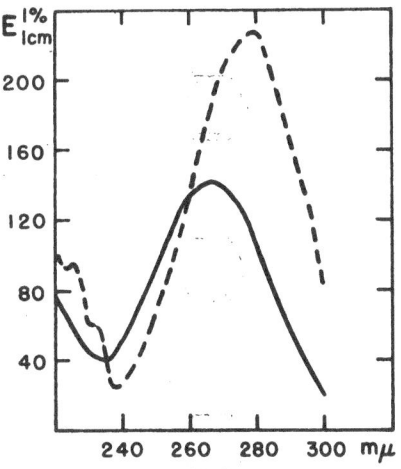

Fig. 64. Extinction curves of 1-phenylbutadiene, in ethanol: ———, *cis* form; and — — —, *trans* form; according to GRUMMIT and CHRISTOPH (*148*). [From: J. Amer. Chem. Soc. **73**, 3479 (1951).]

the sensitivity of the aldehyde group and the rather difficult separation of the isomers. The main product was mostly *cis*-II while *cis*-IV appeared only in traces. The *cis* isomers I and IV have been obtained as chromatographically homogeneous solutions but II and III crystallized; m. p. II, 108°; and III, 176–178°, after sintering at 103°.

When benzene solutions were refluxed in darkness, the all-*trans* compound remained sterically unaltered but under similar conditions 1% and 10%, respectively, of *cis*-II and -III rearranged to the all-*trans* form. The *cis* phenylundecapentaenals showed marked stereolability when exposed to scattered daylight *(Table 26)*. The catalytic effect of iodine, in light, is less important in this set than in most sets already discussed.

No pertinent quantitative data are available for *cis*-1-phenylundecapentaenal that suffers structural alterations during rechromatography and produces a beige-colored zone *above* the all-*trans* zone. The spectral data for this isomer are given with some reservation.

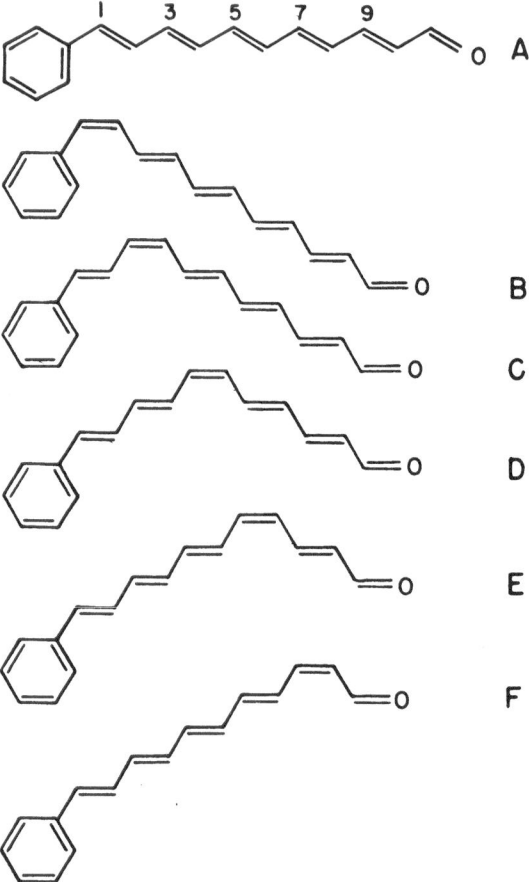

Fig. 65. Models of some stereoisomeric 1-phenylundecapentaenals: *A*, all-*trans*; *B*, *1-cis*; *C*, *3-cis*; *D*, *5-cis*; *E*, *7-cis*; and *F*, *9-cis*.

The spectral characteristics of the stereoisomers appear in *Table 27* and *Figs. 66–68*. In the fundamental band the presence of the aldehyde group causes, as expected, strong dependence of the fine structure on the polarity of the solvent (cf. *436*).

The infrared spectra show a significant steric influence mainly in two regions: (a) in the 7.5–9.0 μ region which is stereochemically sensitive

Table 26. Composition of Some Stereoisomeric Mixtures Obtained from 1-Phenylundecapentaenals (benzene solutions) (*122*).

Steric form treated	Treatment	Content in the recovered substance (%)				Recovery (%)
		all-*trans*	*cis*-I	*cis*-II	*cis*-III	
All-*trans*	Diffuse daylight (4 min.) .	95	1	3	1	99
	Insolation (2 hr.)	62	6	25	7	75
	Irradiation (Photoflood, 2 hr.)	58	6	32	4	90
	Iodine catalysis in light (30 min.)	84	1	5	10	94
	Melting crystals (90 sec.) ..	72	1	27	0	69
cis-II	Diffuse daylight (4 min.) .	12	~0	76	12	99
	Insolation (2 hr.)	61	5	27	7	61
	Irradiation (Photoflood, 2 hr.)	54	2	39	5	78
	Iodine catalysis in light (30 min.)	82	3	5	10	85
	Melting crystals (90 sec.) .	69	3	28	0	60
cis-III	Diffuse daylight (4 min.) .	48	1	13	38	99
	Insolation (20 sec.)	9	0	4	87	99
	Iodine catalysis in light (4 min.)	48	1	13	38	99

Table 27. Some Spectral Characteristics of Stereoisomeric 1-Phenylundecapentaenals (in hexane) (*122*).

Steric form	Position of λ_{max} (mμ)	λ_{max} shift (mμ)	$E_{mol}^{1\ cm} \times 10^{-4}$ at λ_{max}	Fine structure in main band	Position of *cis*-peak (mμ)
All-*trans*......	392	—	10.0	very extensive	—
cis-I	389	3	8.1	extensive	—
cis-II	390	2	7.5	extensive	290
cis-III........	390	2	6.9	extensive	290
cis-IV	387	5	6.6	moderate	—

also in diphenylpolyenes, without the possibility of an interpretation; and (b) in the 10.0–10.6 μ region where a conjugated *cis-trans* diene group causes splitting of the 10.02 μ band present in the all-*trans* spectrum. This doublet formation becomes strongly manifest in the case of *cis*-II (new band at 10.19 μ); but only a small shoulder appears (at 10.13 μ) in the *cis*-III curve (*Fig. 69*, p. 157).

At present the following tentative configurational assignments can be made in this set. Considering the smallness of the λ_{max} shift, the *cis* forms I–III very probably represent mono*cis* isomers.

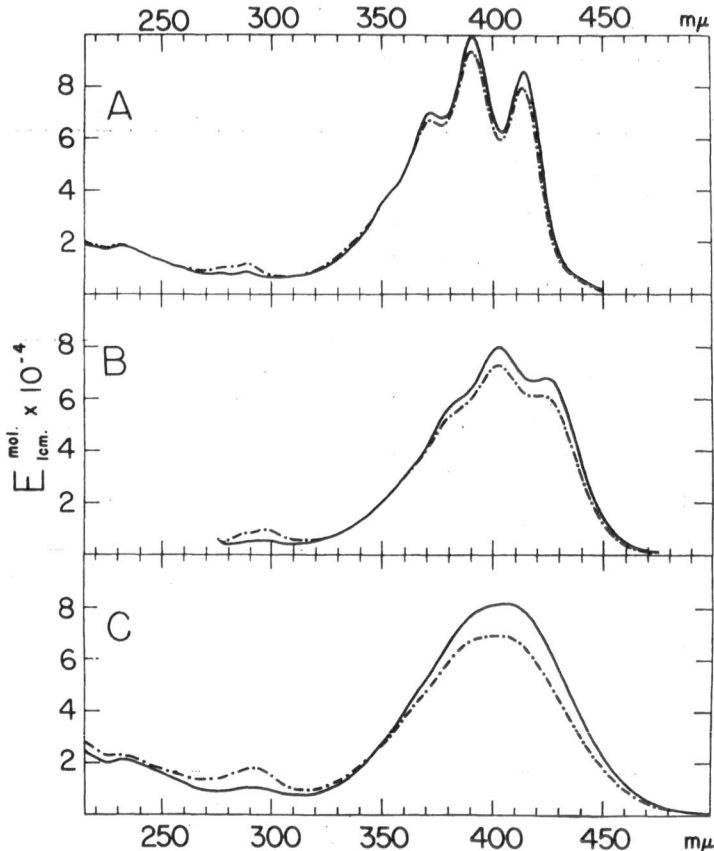

Fig. 66. Molecular extinction curves of all-*trans*-1-phenylundecapentaenal: ————, fresh solution; and
· —·—·—·, mixture of stereoisomers upon catalysis by iodine; *A*, in hexane; *B*, in benzene; and *C*, in
methanol (*122*). [From: J. Amer. Chem. Soc. **79**, 3854 (1957).]

Cis-I, because of its strongest adsorption affinity among the *cis* forms
and the absence of a *cis*-peak may well possess the 9-*cis* configuration
and hence an essentially straight molecular shape (Model F, p. 152).
For *cis*-II or *cis*-III the 9-configuration is excluded because of the presence
of *cis*-peaks; and Model D is excluded because of the moderate height
of these peaks as compared with the maximum value observed in the
diphenyloctatetraene set. Consequently, we propose the 3-*cis* configuration
(Model C) for *cis*-II and the 7-*cis* configuration (E) for *cis*-III (or perhaps
vice versa). The minor isomer *cis*-IV seems to be a di*cis* compound
although, considering the flatness of its curve in the *cis*-peak region and
the almost complete lack of fine structure in the main band, the sterically
hindered 1-mono*cis* configuration (B) cannot be excluded.

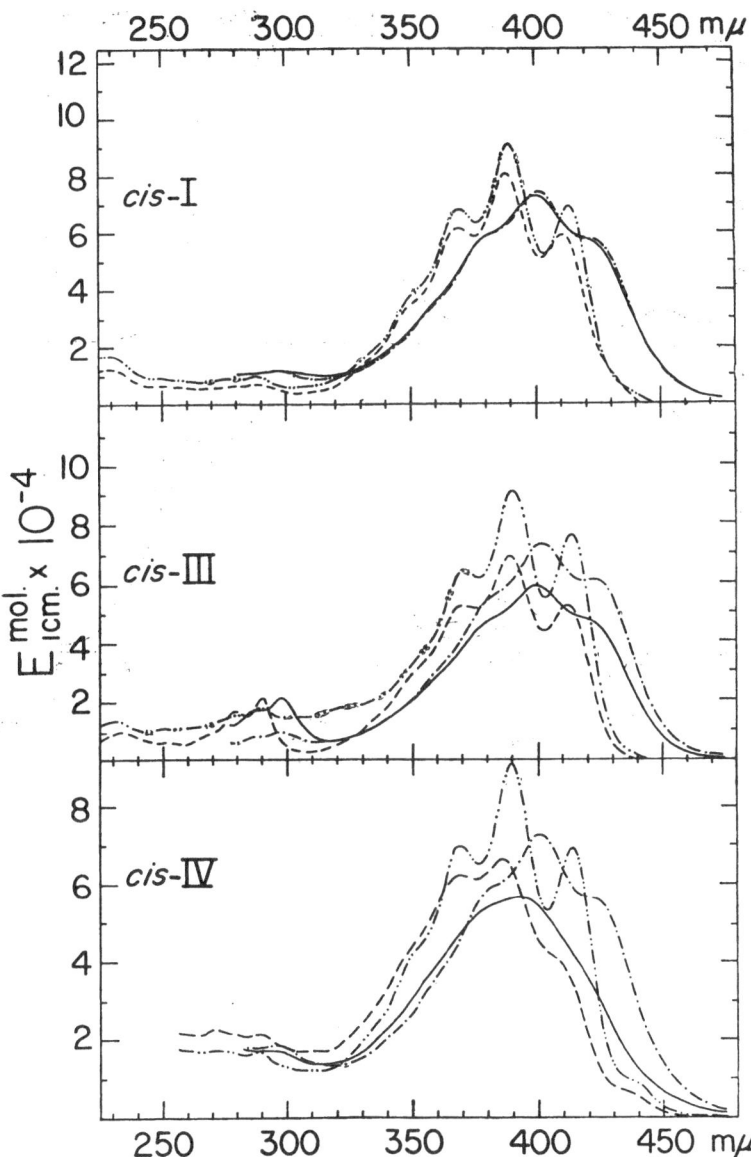

Fig. 67. Molecular extinction curves of some *cis* 1-phenylundecapentaenals: Fresh solutions: — — —, in hexane; and ————, in benzene. Mixture of stereoisomers upon catalysis by iodine: ·· — ·· — ··, in hexane; and · — · — ·, in benzene (*122*). [From: J. Amer. Chem. Soc. **79**, 3854 (1957).]

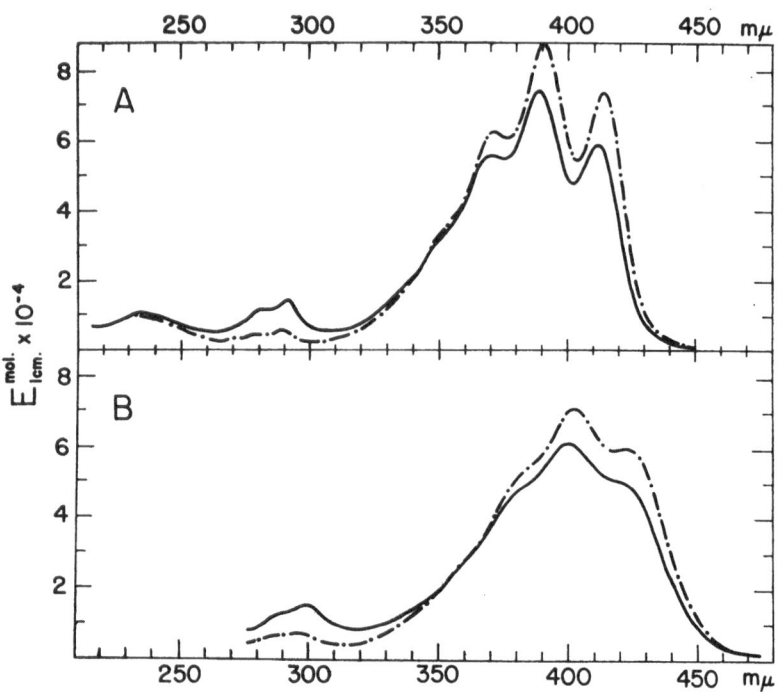

Fig. 68. Molecular extinction curves of *cis*-II-1-phenylundecapentaenal: ―――――, fresh solution; **and** ·—·—·—, mixture of stereoisomers upon catalysis by iodine: *A*, in hexane; and *B*, in benzene (*122*). [From: Amer. Chem. Soc. **79**, 3854 (1957).]

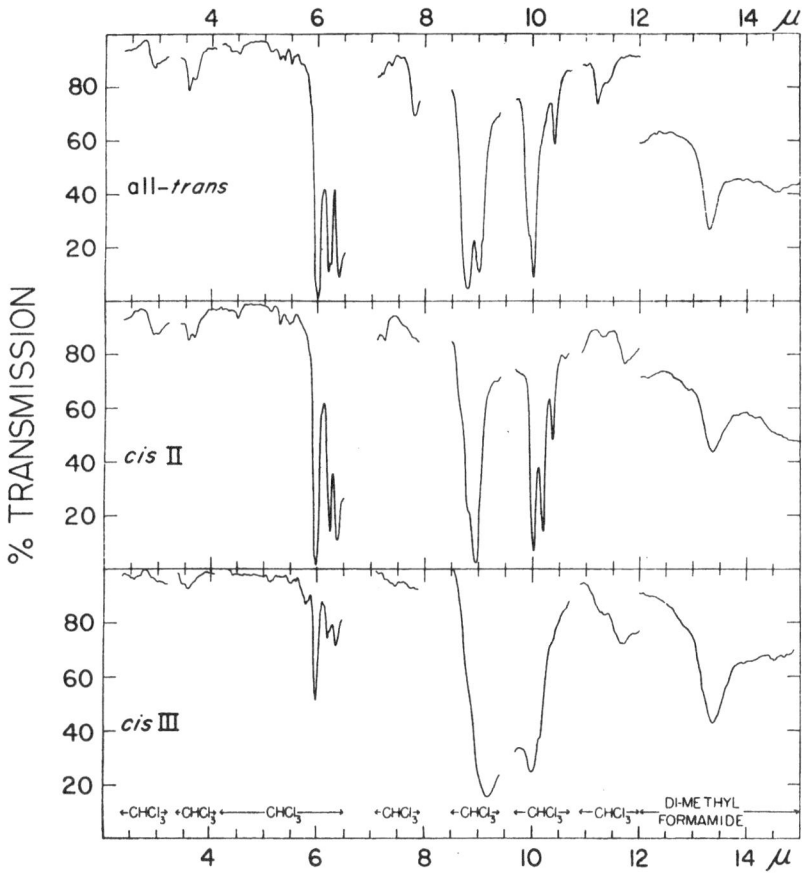

Fig. 69. Infrared spectra of some stereoisomeric 1-phenylundecapentaenals: 2.0–12.0 μ in chloroform, and 12.0–15.0 μ in dimethylformamide (1-mm. cells). Concentrations: all-*trans* and *cis*-II, 0.7%; *cis*-III, 0.07% (microcell) (*122*). [From: J. Amer. Chem. Soc. **79**, 3854 (1957).]

XIV. α,ω-Diphenylpolyenes and Related Compounds.

$$C_6H_5 \cdot (CH=CH)_n \cdot C_6H_5$$

1. General Remarks.

Representatives of this series, up to $n = 8$, have been synthesized by KUHN and WINTERSTEIN long ago (269). In each set they obtained the same end product, even when ten different routes had been followed. Although the all-*trans* configuration had been proved by X-ray studies for the samples $n = 3$ and 4 [HENGSTENBERG and KUHN (160)], the existence of *cis* forms was predicted by KUHN and WINTERSTEIN (270) (p. 8). In 1939 COOK and JONES wrote (71):

> ". . . the behaviour of diphenyl-hexatriene and -octatetraene on a chromatographic column after irradiation in benzene-petroleum ether solution . . . indicates the formation of labile (*cis*-) isomerides."

In 1942–1954 the writer's research group has re-investigated the stereoisomerization of diphenylbutadiene and also described several *cis* forms of diphenylhexatriene and diphenyloctatetraene which were obtained by rearrangement of the all-*trans* compounds. The preparation of the two *cis* diphenylbutadienes upon partial, stereospecific reduction of acetylenic derivatives had been described by STRAUS much earlier (427).

The calculated number of hindered and unhindered *cis* diphenylpolyenes for $n = 1$ to 6 appear in *Table 28* which shows that the hindered configurations are in great majority.

Table 28. Calculated and Observed Number of Sterically Hindered and Unhindered *cis* Isomers in Some Diphenylpolyene Sets.

Stereoisomeric set	Calculated			Observed		
	hindered	unhindered	total	hindered	unhindered	total
Diphenylethylene (stilbene).	1	0	1	1	0	1
Diphenylbutadiene	2	0	2	2	0	2
Diphenylhexatriene	4	1	5	2	1	3*
Diphenyloctatetraene	7	2	9	2	1	3
Diphenyldecapentaene	14	5	19			
Diphenyldodecahexaene....	26	9	35			

* And a trace-isomer.

In practical experimentation exposure to sunshine or to artificial light are much used rearranging methods ($n = 1$ to 4). Catalysis by iodine, in light, is also useful, although, for some reason, the equilibria favor the all-*trans* configuration much more in diphenylpolyene than in carotenoid sets.

Each *cis* diphenylpolyene obtained by our group was adsorbed below the corresponding all-*trans* zone. The location of zones is facilitated

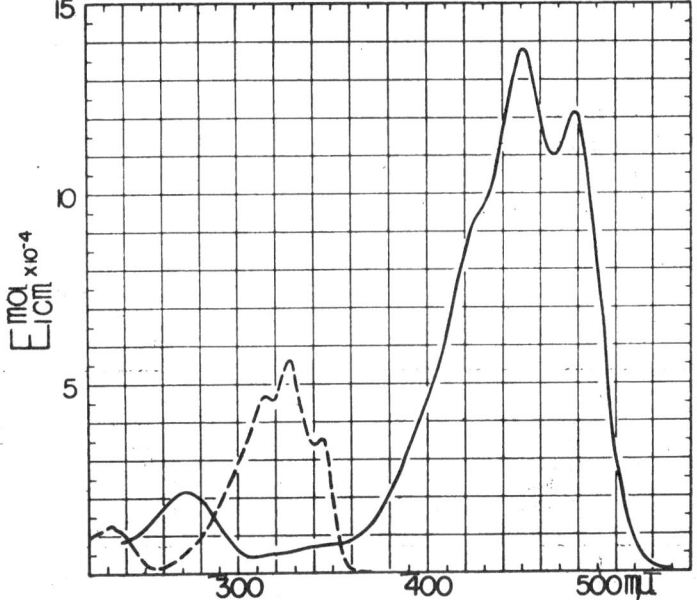

Fig. 70. Comparison of the fundamental bands of — — —, diphenylbutadiene, and ————, β-carotene (in hexane).

by the fluorescence in UV light. *Trans* → *cis* rotations weaken the fluorescence. As would be expected, the fluorescence of 1,1,4,4-tetra-phenylbutadiene, which is incapable of steric rearrangement, is unaffected by illumination (*162, 163*).

Ultraviolet Spectra. Most of the regularities observed in the field of carotenoids are also valid in diphenylpolyene sets. The characteristic differences in the position and height of the fundamental band are illustrated in *Fig. 70*.

Upon cooling a diphenyloctatetraene solution to $- 196°$, the extinction values increase substantially, the fine structure is accentuated, and new maxima appear [HAUSSER, KUHN and SEITZ (*153*)].

The λ_{max} shifts observed in *cis* diphenylpolyenes are considerable; in the homologous series, $C_6H_5 \cdot (CH=CH)_n \cdot C_6H_5$, they decrease with increasing *n*. As in carotenoids, central-mono*cis* configurations cause abnormally low values. If $n > 2$ and the molecule is bent, fine structure appears in the *cis*-peak region.

Infrared Spectra. The following regions were found to be stereochemically sensitive *(Table 29)* [LUNDE and the writer *(291)*].

(a) The 7.00–7.09 μ region: in-plane bending vibration of H atoms attached to *cis* double bonded carbons. This band was absent from all-*trans* curves. An exceptionally high wavelength position, 7.31 μ, was found for di*cis*-diphenylbutadiene, the only all-(hindered)-*cis* form studied in this series.

(b) The 10.0–10.6 μ region: out-of-plane rocking of H atoms attached to a *trans* double bond within a conjugated *cis-trans* diene group. Evidently, the presence of a *trans* double bond is necessary to bring about this maximum, and hence it is understandable that the band is absent from

Table 29. Stereochemically Significant Infrared Bands of Some Diphenylpolyenes *(291)*. (The figures in parentheses represent units of 10% absorption, e. g. (6) stands for 60% absorption; m = medium, ms = medium strong, and s = strong band; wavelengths in μ.)

Stereo-isomeric set	Con-figuration	In-plane bending vibration of H atoms attached to *cis* double bonded carbons	Out-of-plane rocking of H atoms attached to a *trans* double bonded carbon, in conjugation with		Out-of-plane rocking of H atoms attached to *cis* double bonded carbons	Out-of-plane rocking of a pair of equivalent H atoms in the rings	
			any C=C bond	a *cis* C=C bond		hindered	unhindered
Stilbene	*trans* ...	—	10.42(9)	—	—	—	14.48(9)
	cis	7.09(2)	—	—	12.84(8)	14.37(9)	
Diphenyl-butadiene	all-*trans*.	—	10.15(9)	—	—	—	14.54(9)
	mono*cis*.	7.07(3)	10.14(6)	10.58(6)	12.90(5)	14.33(8)	14.50(8)
	di*cis*....	7.31(2)	—	—	12.91(7)	14.20(8)	14.39(9)
Diphenyl-hexa-triene	all-*trans*.	—	10.06(8)	—	—	—	14.50(5)
	3-*cis*....	7.03(2)	10.06(5)	10.42(10)	12.85(2)	—	14.50(9)
	1-*cis*....	7.05(2)	10.06(9)	10.42(5)	12.93(8)	14.26(9)	14.50(9)
Diphenyl-octa-tetraene	all-*trans*.	—	9.90(9)	—	—	—	14.44(5)
	3-*cis*....	7.04(2)	10.11(8)	10.30(9)	12.91(4)	—	14.52(9)
	1-*cis*....	7.05(3)	10.05(10)	10.30(7)	12.95(7)	14.30(9)	14.51(9)
	1,3-di*cis* or 1,5-di*cis*	7.00(2)	10.10(8)	10.30(8)	12.95(8)	14.15(7)	14.49(9)
	3,5-di*cis**	7.28 m	10.15 ms	10.55 s	12.85 m		

* Synthesized by AKHTAR, RICHARDS and WEEDON *(4)*.

the cis-stilbene and dicis-diphenylbutadiene curves. Furthermore, in contrast to all-trans forms, each isomer that contains conjugated cis and trans double bonds shows in this region a doublet in which the relative intensities depend on the position of the cis double bond. In the case of terminal position, i. e. conjugation with only one aliphatic trans double bond, the shorter wavelength maximum of the doublet shows higher intensity (cf. terminal cis-diphenylhexatriene or -octatetraene). The relative intensities are, however, reverted when the cis double bond is conjugated with two trans double bonds.

(c) The 12.84–12.95 μ region: out-of-plane rocking of H atoms attached to cis double bonded carbons.

(d) The 14.5 μ region: the aromatic band was split in some instances and yielded a new band at shorter wavelengths (14.15–14.37 μ). In conformity with the UV spectra this effect has been assigned to the presence of terminal cis double bond(s), i. e. to hindered cis configuration.

Although much remains to be done in this field, the configurations proposed in this Chapter are in accordance with the available IR readings.

Recently, STEGEMEYER (413) has reported that the 3.46 μ band observed by us in the cis-stilbene spectrum (291) was due to a contamination by photochemically formed tetraphenylcyclobutane. Hence, a re-investigation of this region is desirable (294).

2. Stilbene Set.

The extensive stilbene literature (n = 1) will not be reviewed here in detail.

cis-Stilbene (Fig. 71) was detected by OTTO and STOFFEL (343) in 1897 by the interaction of isostilbene dibromide and sodium thiophenolate;

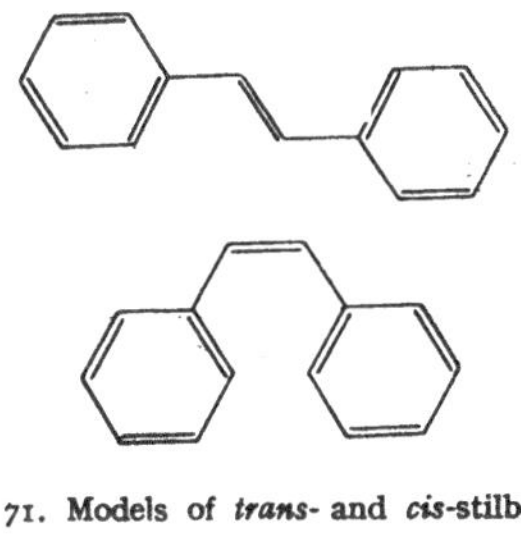

Fig. 71. Models of trans- and cis-stilbenes.

later it was obtained by STRAUS (427) from diphenylacetylene. A practical method of preparation is decarboxylation of α-phenylcinnamic acid (434). cis-Stilbene (m. p. 1°) can also be obtained by the irradiation of ordinary (trans) stilbene (m. p. 124°) in ultraviolet light (415, 285, 408, 518).

cis-Stilbene is sterically hindered, and in this limiting case both conflicting hydrogens are attached to benzene rings (Fig. 72). The difference in optical properties of cis- and trans-stilbene has been interpreted by LEWIS et al. (284, 285) on the assumption that the repulsion of two o-hydrogens forces the molecule out of plane and diminishes resonance. In contrast, the all-*trans*-stilbene molecule is strainless (217).

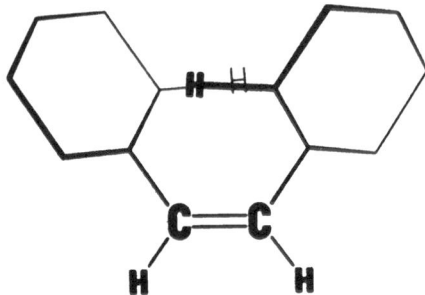

Fig. 72. Overlapping of hydrogen atoms in *cis*-stilbene; according to LEWIS et al. (285). [From: J. Amer. Chem. Soc. 62, 2973 (1940).]

For the separation of *cis*- and *trans*-stilbenes use has been made of fractional crystallization, distillation, etc. The stereoisomeric mixture can also be resolved chromatographically, upon painting a permanganate streak along the column; where the reagent crosses either stilbene zone, it turns brown (518).

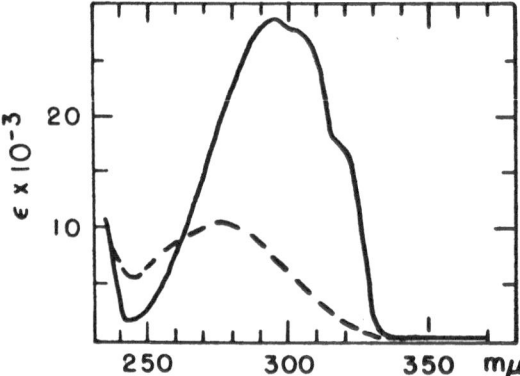

Fig. 73. Molecular extinction curve in abs. alcohol: ————, *trans*-stilbene; and — — —, *cis*-stilbene; according to CALVIN and ALTER (50). [From: J. Chem. Physics 19, 765 (1951).]

The spectral curves of the two stilbenes appear in *Fig. 73*. The λ_{max} shift amounts to 16 mμ (in alcohol). A discussion of the stilbene spectra in comparison with those of chain molecules consisting of alternating benzene rings and ethylenic bonds was presented by DALE (83).

For the spectra of some other *cis trans* pairs see e. g. (387 a).

cis-Stilbene is more thermostable than might have been expected: in the melt, at 214°, only 8% undergoes rearrangement in 20 hours

[TAYLOR and MURRAY *(435)*]. For the kinetics of similar processes, cf. e. g. KISTIAKOWSKY and SMITH *(247)*, CALVIN and ALTER *(50, 51)*, MAYO and WALLING *(308)*. *cis*-Stilbene is rearranged by iodine, in light. According to KHARASCH et al. *(246)* it is not isomerized by HBr in darkness, either in the presence or absence of air; peroxidic agents, however, cause a rapid *cis → trans* change.

For the influence of ferromagnetic metals cf. *(441, 442)*. Rearrangement by N_2O_4, *(403)*.
Combustion and ozonization heats of the two stilbenes: *(42)*. — Use of nuclear magnetic resonance to determine configurations of stilbenes: *(78)*. — Spectroscopic study of *cis* and *trans* dibenzoylethylenes: *(257)*. — Quantum yields of the photochemical *cis ⇄ trans* rearrangements of *p,p'*-nitromethoxystilbene in various solvents: *(399)*.

3. Diphenylbutadiene Set.

This compound ($n = 2$) may assume an all-*trans*, a hindered mono*cis*, and a hindered di*cis* configuration *(Fig. 74)*.

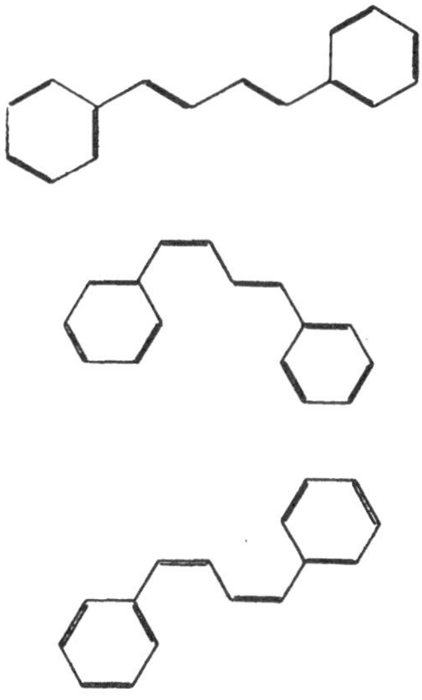

Fig. 74. Models of stereoisomeric diphenylbutadienes: top, all-*trans*; middle, *cis-trans*; and bottom, *cis-cis*.

As mentioned, STRAUS *(427)* must be credited with the first preparation of the two *cis* isomers. As early as 1905 he treated diphenyldiacetylene, $C_6H_5 \cdot C\equiv C—C\equiv C \cdot C_6H_5$, with copper-coated zinc and obtained a relatively labile but crystallizable isomer that on standing or, more rapidly, in sunlight yielded ordinary (all-*trans*) diphenylbutadiene. When the reduction process was prematurely interrupted and the isolated diphenylbutenine, $C_6H_5 \cdot C\equiv C—CH=CH \cdot C_6H_5$, further reduced, a second *cis* isomer of oily nature appeared. The *cis* forms can also be obtained by reduction with hydrogen over Pd/BaSO$_4$ (*242, 341*).

A re-investigation of the diphenylbutadienes was carried out in collaboration with SANDOVAL, PINCKARD and WILLE (*395, 355*) by means of chromatographic and fluorescence methods. The STRAUS data concerning the crystalline *cis* isomer were confirmed but the oily isomer revealed on the column the presence of impurities. The reduction of diphenyldiacetylene with hydrogen over Pd/BaSO$_4$, yielded both *cis* isomers. A simple route for the preparation of the mono*cis* form is exposure of all-*trans* solutions to a projection lamp for a day; the ratio in the recovered pigment was, *trans* : *cis* = 16 : 84.

The chromatographic top-to-bottom sequence on alumina is, *trans-trans-*, *cis-trans-*, and *cis-cis-*diphenylbutadiene.

The all-*trans* zone can be located by its intense bluish fluorescence in ultraviolet light, and the *cis* zones by their partial quenching of the weak fluorescence of commercial alumina. A sharper separation of the *cis-trans-* and *cis-cis-*diphenyl-butadienes was accomplished as follows: While the all-*trans* zone remained near the top, the *cis* zones were washed fractionally with petroleum ether into the filtrate. A small sample of each fraction was collected and, after the addition of a drop of iodine solution in petroleum ether and brief illumination with an electric bulb (to give the all-*trans* form), each sample was tested for fluorescence in UV light, under a black cloth. While the di*cis* form was passing through, this test gave a positive result, then it became negative (empty interzone), and finally positive again when the *cis-trans* form began to run into the filtrate.

The three chromatographically homogeneous preparations had the following properties:

trans-trans form: m. p. 152–153°, λ_{max} at 328 mμ (in hexane), fluorescent;
cis-trans form: oil, λ_{max} at 313 mμ, not fluorescent; and
cis-cis form: m. p. 70.5°, λ_{max} at 299 mμ, not fluorescent.

The λ_{max} shift for one *trans* → *cis* rotation (14–15 mμ in hexane) is here much larger than that observed in the carotenoid class but only slightly different from the corresponding value for stilbene (16 mμ).

As *Fig. 75* shows, both *cis* diphenylbutadienes have degraded spectra, very different from the all-*trans* curve. Evidently, as in the carotenoid series, a single shift, *trans* → hindered *cis*, eliminates the fine structure and decreases the maximum extinction to about half of the initial value. Upon iodine catalysis of the mono*cis* compound, in light, the fine structure and the high extinction values are recovered because the resulting stereo-isomeric equilibrium contains 97% all-*trans* form. The effect is still more impressive when di*cis*-diphenylbutadiene is catalyzed; then the increase of the extinguished area may well be compared with that observed during the conversion, poly*cis* lycopene → all-*trans*-lycopene (p. 7 .).

Although no clearly differentiated *cis*-peak appears in the compressed curve of the slightly bent mono*cis*-diphenylbutadiene, it should not be overlooked that the extinction at 230–240 mμ is the highest of the three stereoisomers; moreover, that the maximum in this region is located at a 9 mμ longer wavelength than the corresponding maximum of the all-*trans* curve, indicating a "hidden" *cis*-peak.

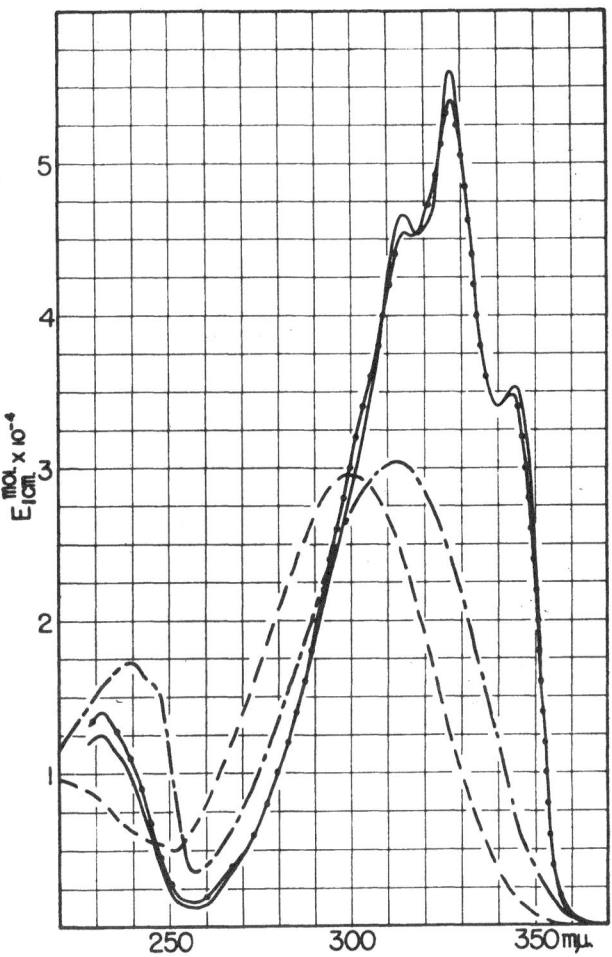

Fig. 75. Molecular extinction curves of the three stereoisomeric diphenylbutadienes, in hexane: ———— *trans-trans*; —·—, *cis-trans*; — — —, *cis-cis*; and —·—·—·—, mixture of stereoisomers after iodine catalysis of any of the three forms (355). [From: J. Amer. Chem. Soc. 70, 1938 (1948).]

With reference to relative stabilities of the diphenylbutadienes the following statements can be made.

The configurations of all three isomers are light-sensitive to a different degree, depending on the conditions. In a first (chance) experiment, a very dilute solution of all-*trans*-diphenylbutadiene was left standing on the window sill, in diffuse light, for half an hour; the fine structure disappeared from the fundamental band and the extinction at λ_{max} was reduced by about one half. Since practically the original curve was

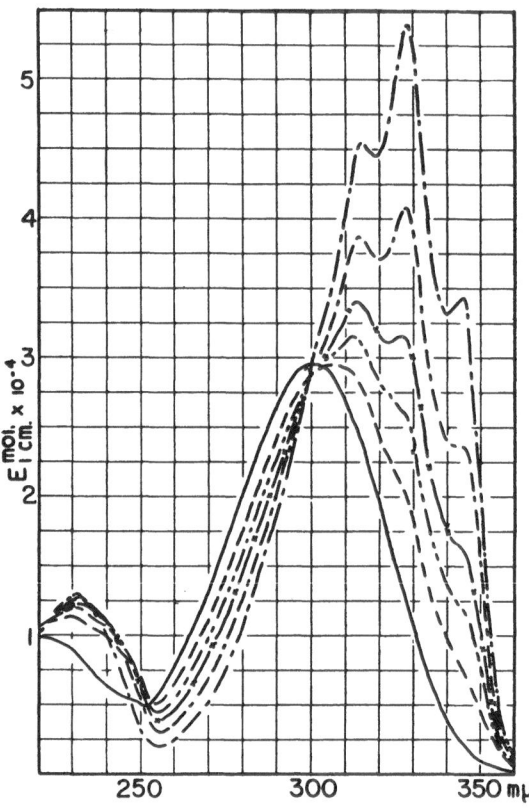

Fig. 76. Influence of illumination on the stereoisomerization of *cis-cis*-diphenylbutadiene caused by iodine catalysis: ————, fresh hexane solution, with iodine, kept in darkness for 30 min.; — — —, after 80 sec.; — · · —, 140 sec.; — x — x —, 200 sec.; — — · — —, 320 sec.; and — · — ·, 1500 sec. illumination (355).
[From: J. Amer. Chem. Soc. 70, 1938 (1948).]

recovered on addition of iodine, no irreversible destruction had taken place.

Under these conditions the main process is a stepwise rearrangement, *cis-cis* → *cis-trans* → *trans-trans*, as illustrated in *Fig. 76*. If direct *cis-cis* → *trans-trans* isomerization had taken place, it would have caused the appearance of fine structure during early stages of the illumination. This, however, was not the case as demonstrated by the extinction curve taken after 80 sec. exposure. (The curve of an artificial mixture, 95%

cis-cis + 5% all-*trans*, showed bulges at 328 mμ and 344 mμ, like the 140 sec. curve in Fig. 76.)

In the absence of iodine not the all-*trans* but the mono*cis* form is the most photostable one; and the two other isomers tend to assume

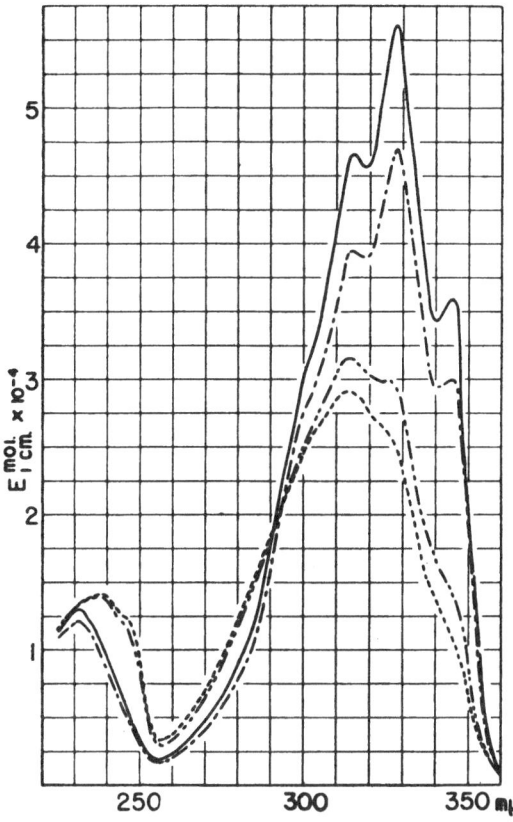

Fig. 77. Molecular extinction curve of *trans-trans*-diphenylbutadiene, in hexane, and its shift toward the curve of the *cis-trans* form during exposure to sunshine: ————, fresh solution; — ·· —, after 1 min.; — — —, after 10 min. insolation; and — · — ·. after iodine catalysis at the end of a 10 min. insolation period (355). [From: J. Amer. Chem. Soc. 70, 1938 (1948).]

this configuration in sunshine or in the light of an electric bulb (*Figs.* 77 and 78).

In darkness, the solutions of the three diphenylbutadienes are remarkably thermostable; the individual configurations can be preserved for several months at 4°. They are but little altered by an hour's refluxing or keeping in the molten state at 205° for a few minutes. In the absence of solvents the oily *cis-trans* form has, however, the tendency to rearrange on long standing at room temperature, even in near-darkness, and to deposit crystals of the all-*trans* form. In strong light such crystals appear within a minute. As we will see, the oily, hindered *cis* forms of diphenyl-hexatriene and -octatetraene behave similarly (pp. 170, 177).

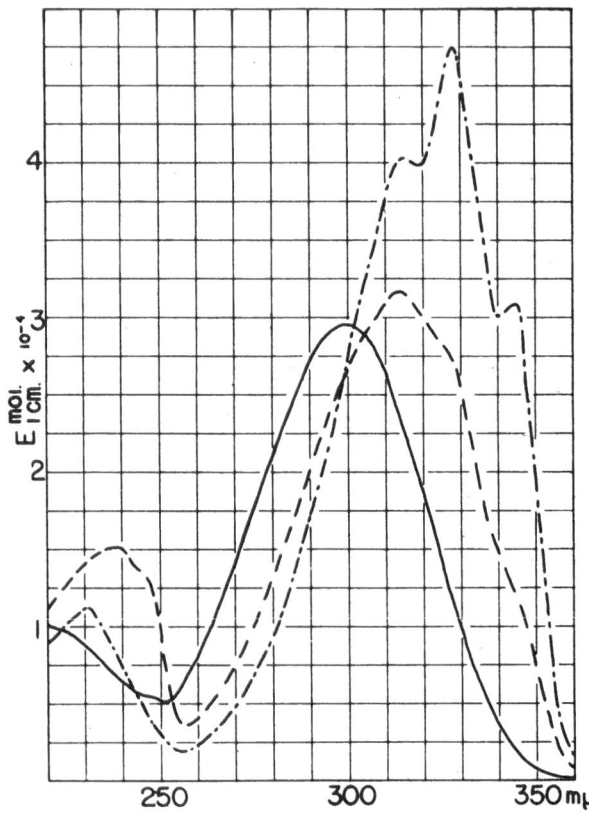

Fig. 78. Molecular extinction curve of *cis-cis*-diphenylbutadiene, in hexane, and its shift toward the curve of the *cis-trans* form during exposure to sunshine: ————, fresh solution; — — —, after 10 min. insolation; and · — · — ·, after iodine catalysis at the end of a 10 min. insolation period (355). [From: J. Amer. Chem. Soc. 70, 1938 (1948).]

To summarize: although the relative stabilities of the three stereo-isomeric diphenylbutadienes are subject to considerable variation, under no condition applied was the *cis-cis* form the most stable one.

Diphenylbutadiene can be considered as the first member of the series, $H \cdot (C_6H_4 \cdot CH=CH—CH=CH)_n \cdot C_6H_5$. On irradiation, the next two members do show the *cis*-peak effect, but pure *cis*-polyphenyl-polybutadienes have not yet been obtained [DREFAHL and PLÖTNER (99)].

4. Diphenylhexatriene Set.

This compound ($n = 3$) may assume five *cis* configurations of which only the central-*cis* (3-*cis*) is sterically unhindered *(Fig. 79)*.

Because all-*trans*-diphenylhexatriene yields only traces of *cis* isomers on refluxing, the method of choice is catalysis by iodine *(Fig. 80)* or

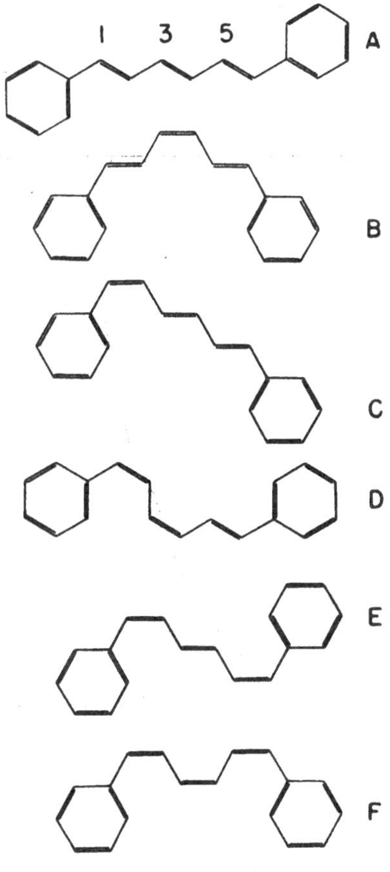

Fig. 79. Models of the six stereoisomeric 1,6-diphenylhexatrienes: *A*, all-*trans*; *B*, 3-mono*cis* (central-*cis*); *C*, 1-mono*cis* (hindered); *D*, 1,3-di*cis* (hindered); *E*, 1,5-di*cis* (hindered); and *F*, all-*cis* (hindered).

exposure to sunshine. Thus, LUNDE and the writer (*292*) have observed four isomers, termed *cis* forms I–IV, in the sequence of decreasing adsorption affinities; *cis*-IV appears only in traces (*Table 30*, p. 170). *Cis*-I was crystallized, m. p. 107° (versus 203° of the all-*trans* compound), but II and III were isolated only as chromatographically homogeneous oils that fluoresce in UV light. Their relative stabilities can be described as follows.

Table 30. Composition of Some Stereoisomeric Mixtures Obtained from all-*trans*-Diphenylhexatriene.

Treatment	Content in the recovered substance (%)				Recovery (%)
	unchanged all-*trans*	*cis*-I	*cis*-II	*cis*-III	
Iodine catalysis	94.1	1.8	3.8	0.3	99
Insolation (60 min.)........	85.6	4.6	9.1	0.7	72
Refluxing (hexane)	only traces of *cis* forms present				~ 100
Melting crystals (15 min.) ..	78	4	17	1	13

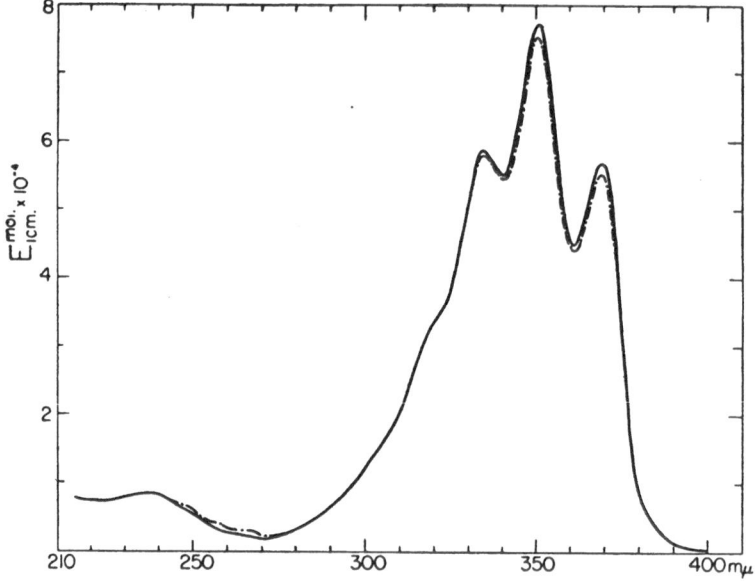

Fig. 80. Molecular extinction curve of diphenylhexatriene in hexane: ———, fresh solution of the all-*trans* form; and ·—·—·, mixture of stereoisomers after 20 min. illumination in the presence of iodine (94% unchanged) (*292*). [From: J. Amer. Chem. Soc. 76, 2308 (1954).]

In solution, in darkness, all-*trans*-diphenylhexatriene is the most thermostable form, but it is light-sensitive to a limited extent. When refluxed in the dark, *cis*-I is less resistant than the all-*trans* and *cis*-II forms. In scattered daylight the stability of *cis*-I is well comparable with that of the all-*trans* compound; *cis*-II is much more photosensitive. In the absence of solvents a few minutes exposure of oily *cis*-II samples to sunshine resulted in the deposition of all-*trans* crystals (same behavior as that of mono*cis*-diphenylbutadiene, p. 16). *Cis*-III-diphenylhexatriene shares the increased thermolability with *cis*-I and the relatively high degree of photolability with *cis*-II. When during chromatographic

Table 31. Some Spectral Characteristics of Stereoisomeric
Diphenylhexatrienes (in hexane).

Configuration	Position of λ_{max} (mμ)	$E_{1\ cm.}^{mol.} \times 10^{-4}$ at λ_{max}	Fine structure in main band	λ_{max} shift (mμ)	Position of maxima in the cis-peak region (mμ)*	$E_{1\ cm.}^{mol} \times 10^{-4}$ at cis-peak
All-*trans*	351	7.72	sharp	0	none	0
Central-*cis* (cis-I)	349–350	5.15	moderate	1–2	260, 268	2.0
I-*cis* (cis-II)	336–337	4.87	none	14–15	256, 261–262	1.4
1,3-di*cis* (cis-III)	334–335	4.73	none	16–17	none	0

* This set shows the exceptional feature that the wavelength location of the cis-peak is subject to variation. ·

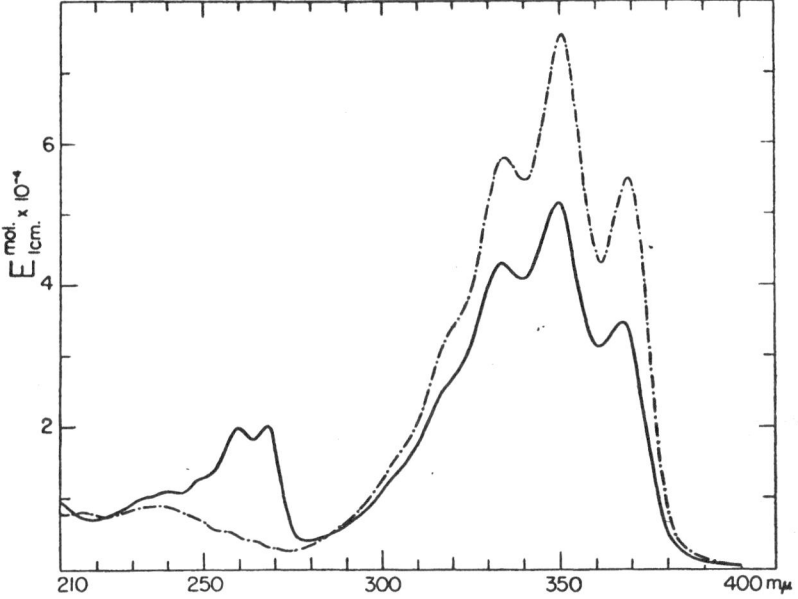

Fig. 81. Molecular extinction curves of central-*cis*-diphenylhexatriene (*cis*-I) in hexane: ————, fresh solution; and · —· —·, mixture of stereoisomers after 60 min. illumination in the presence of iodine (292).
[From: J. Amer. Chem. Soc. 76, 2308 (1954).]

development a *cis*-II or -III zone is exposed to UV light from a short distance, the formation of a new all-*trans* zone is observed on the column surface within a minute or two.

The spectral curves of the three *cis* diphenylhexatrienes show impressive differences: *cis*-I, fine structure in the main band, *cis*-peak present; *cis*-II, no fine structure in the main band, *cis*-peak present; and *cis*-III, no fine structure in the main band, no *cis*-peak present (*Table 31*, p. 171; *Figs. 81* to *83*).

We have assigned the central-*cis* (3-*cis*) configuration to the *cis*-I isomer (Model B, Fig. 79) on the basis of the high *cis*-peak, the fine structure in the fundamental band and the very small λ_{max} shift (1–2 mμ) (the latter feature is also shown by central-mono*cis* carotenoids, p. 78). To *cis*-II the slightly bent, sterically hindered 1-*cis* configuration (Model C)

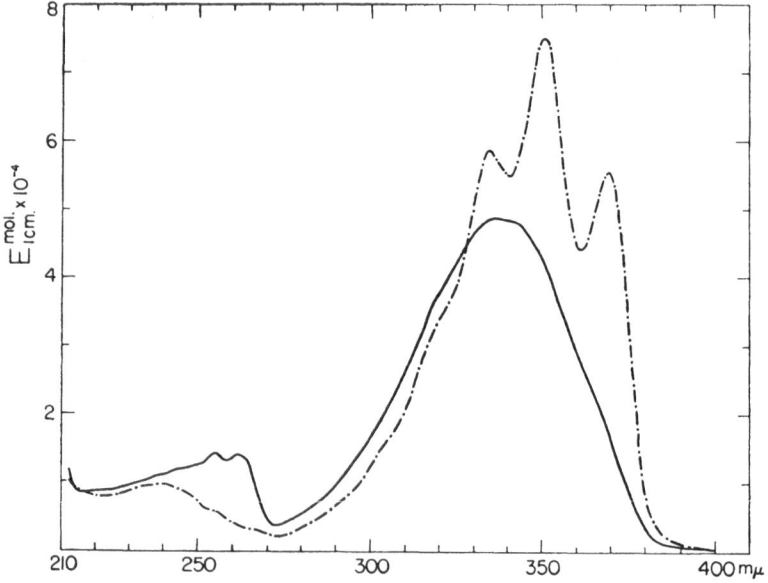

Fig. 82. Molecular extinction curves of 1-*cis*-diphenylhexatriene (*cis*-II) in hexane: ————, fresh solution; and · — · — ·, mixture of stereoisomers after 5 min. illumination in the presence of iodine *(292)*. [From: J. Amer. Chem. Soc. 76, 2308 (1954).]

has been assigned, in accordance with the presence of a moderately high *cis*-peak, lack of fine structure in the main band, and a λ_{max} shift of 14–15 mμ which equals the value found in the diphenylbutadiene set.

For *cis*-III the following considerations have led to the 1,3-di*cis* assignment (Model D): Although each of the remaining configurations, D, E, and F (Fig. 79), could a priori satisfy the outstanding features of the spectrum, a selection became possible by observing the behavior of the isomers in iodine catalysis *(Fig. 84)*. Under the conditions applied the equilibrium, when starting from *cis*-I (hindered), was reached in 3 minutes, but from central-*cis* (unhindered) in 60 minutes. We had postulated that in the simultaneous presence of a hindered and an unhindered *cis* double bond, either *cis* double bond will rearrange approximately at its own rates (cf. p. 132). The sequence of the partial

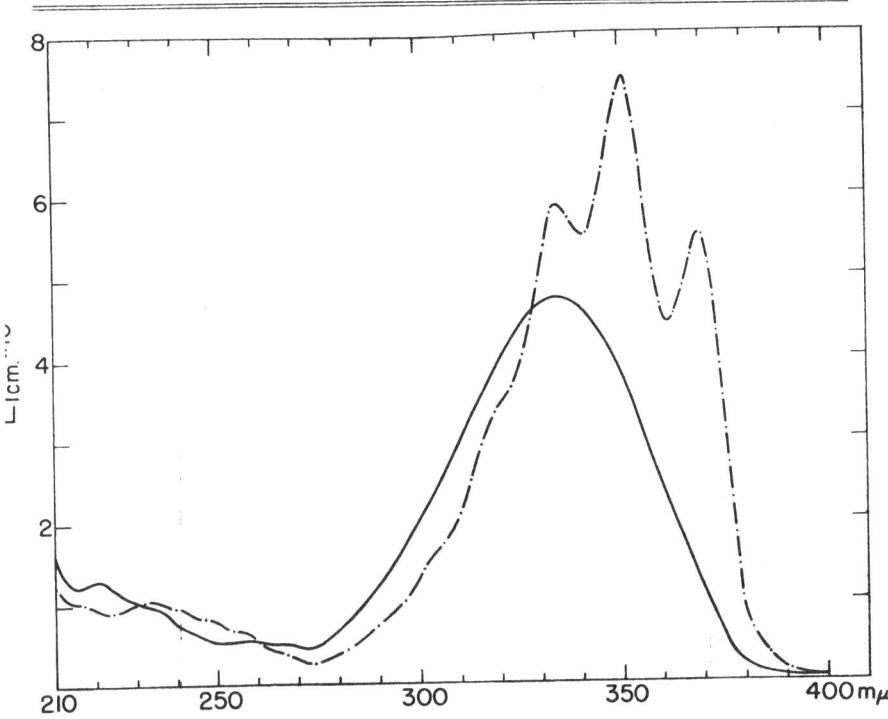

Fig. 83. Molecular extinction curves of 1,3-di*cis*-diphenylhexatriene (*cis*-III) in hexane: ————, fresh solution; and · — — ·, mixture of stereoisomers after 60 min. illumination in the presence of iodine (*292*). [From: J. Amer. Chem. Soc. 76, 2308 (1954).]

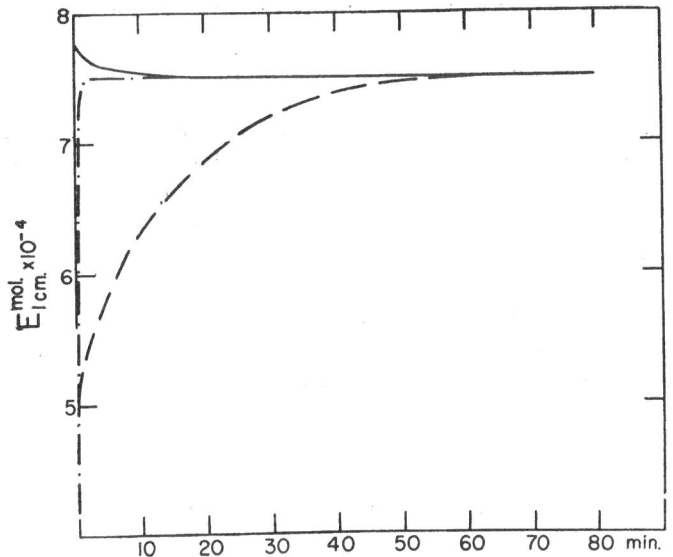

Fig. 84. Change of the extinction value at 351 mμ (λ_{max} of the all-*trans* form, in hexane) during iodine catalysis, in light, starting from the following diphenylhexatrienes: ————, all-*trans*; — — — central-*cis* (*cis*-I); and · — · — ·, 1-*cis* (*cis*-II) (*292*). [From: J. Amer. Chem. Soc. 76, 2308 (1954).]

processes will then be, hindered-unhindered di*cis* → unhindered mono-*cis* → equilibrium mixture (containing much all-*trans* but no hindered *cis* form). This means that on catalysis by iodine the extinction value at λ_{max} would first rise very rapidly but then, when the *cis* → *trans* rotation of the hindered *cis* double bond had come practically to an end, a much more flattened section of the curve would follow. *Cis*-III showed exactly this behavior *(Fig. 85)*. Furthermore, when the catalytic process

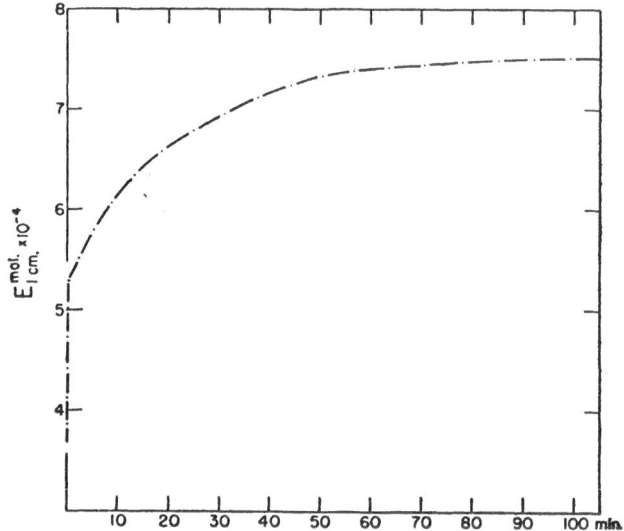

Fig. 85. Change of the extinction value at 351 mμ (in hexane) during iodine catalysis, in light, of *cis*-III-diphenylhexatriene (*292*). [From: J. Amer. Chem. Soc. **76**, 2308 (1954).]

was interrupted after 40 seconds (by switching off the light) and the solution chromatographed without delay, in darkness, the main zone contained 90% central-*cis*-diphenylhexatriene. Hence, the *cis*-III molecule must have *cis* configurations at the central double bond and at least one of the two hindered double bonds. This eliminates Model E (p. 16) that contains the central double bond in *trans*. The choice between the remaining Models D and F was made in favor of D for the following two reasons: (a) the λ_{max} shift for the tri*cis* form F should much exceed the observed value of 16 mμ; and (b) as shown in the carotenoid class, there is no tendency for the formation of all-*cis* configurations in the presence of iodine.

PARA and FORSTER (*344 a*) who have measured the rates of the *cis*-I → *trans* and *cis*-II → *trans* thermal rearrangements in the diphenylhexatriene set have proposed different assignments (1,5-*dicis* for *cis*-III).

Low-melting isomers of some *substituted* diphenylhexatriene derivatives, (*314*).

5. Diphenyloctatetraene Set.

In this set ($n = 4$) the number of possible *cis* isomers is nine of which seven are sterically hindered and only two unhindered, viz. the middle-mono*cis* (3-*cis*) and the middle di*cis* (3,5-di*cis*) forms; some models appear in *Fig. 86*. The diphenyloctatetraenes were investigated in collaboration with LeRosen and Pinckard (*514, 522, 513*).

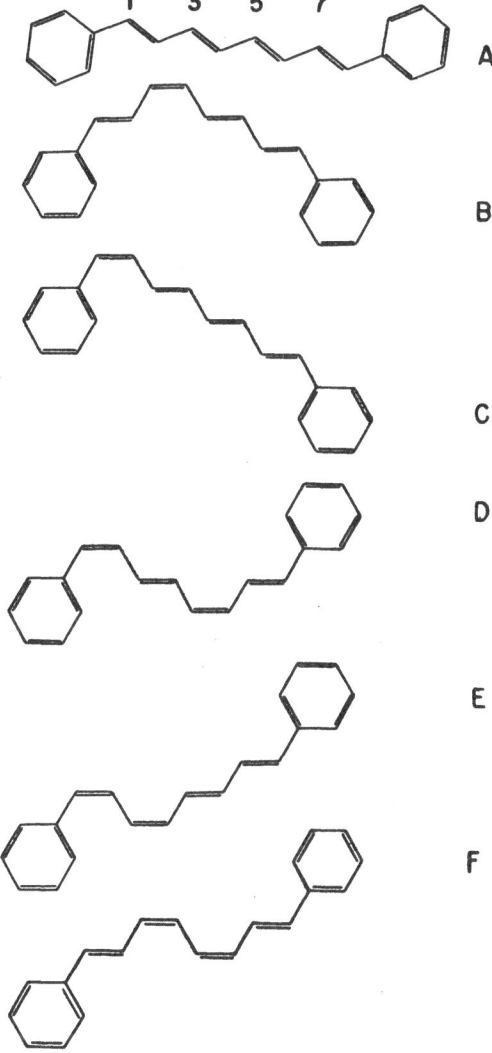

Fig. 86. Models of some 1,8-diphenyloctatetraenes: *A*. a'l-*trans*: *B*, 3-mono*cis*; *C*, 1-mono*cis*; *D*, 1,5-di*cis*; *E*, 1,3-di*cis*; and *F*, 3,5-di*cis*.

When submitted to the treatments listed in *Table 32*, all-*trans*-diphenyloctatetraene (m. p. 235°) yielded three fluorescent, chromato-graphically separable new isomers, termed, in the sequence of decreasing adsorption affinities, *cis*-I (crystalline), *cis*-II (microcrystalline), and *cis*-III (oily). *Cis*-I melts at 139–141°, resolidifies, and melts again, at ~ 235–237°, near the m. p. of the all-*trans* compound. Likewise, *cis*-II shows a double melting point (unsharp at ~ 70–80°, and 190°).

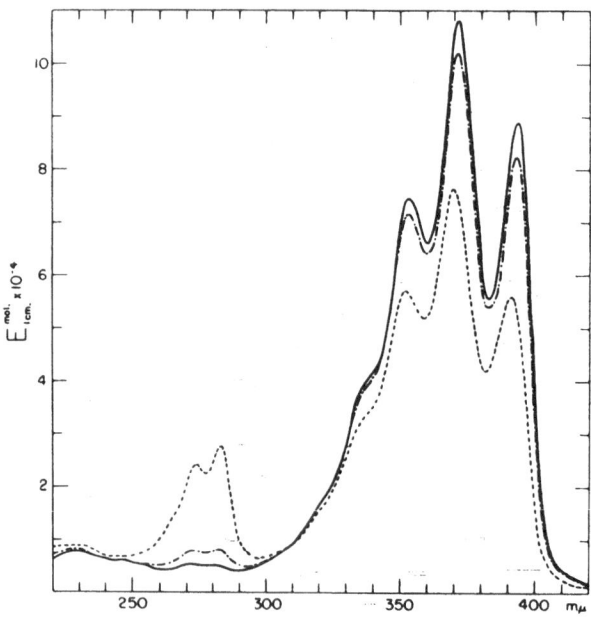

Fig. 87. Molecular extinction curves of all-*trans*- and *cis*-I-diphenyloctatetraenes in hexane: ————, fresh solution of all-*trans*; ......, fresh solution of *cis*-I; and ·—·—·, mixture of stereoisomers after 15 min. iodine catalysis of either form (*522*). [From: J. Amer. Chem. Soc. 76, 4144 (1954).]

Table 32. Composition of Some Stereoisomeric Mixtures Obtained from all-*trans*-Diphenyloctatetraene.

Treatment	Content in the recovered substance (%)				Recovery (%)
	unchanged all-*trans*	*cis*-I	*cis*-II	*cis*-III	
Iodine catalysis (15 min.)	86.2	11.1	2.2	0.5	99
Insolation (1 hr.)	88.8	8.6	2.2	0.4	76
Illumination (Photoflood) (1 hr.)	88.5	9.7	1.5	0.3	96
Refluxing (benzene) (30 min.)........	90.8	8.0	1.0	0.2	96
Melting crystals (5 min.)	77.3	22.0	0.7	—	20

All-*trans*-diphenyloctatetraene, as all-*trans*-diphenylhexatriene, is more thermostable in solution than any of its *cis* isomers, although it is somewhat

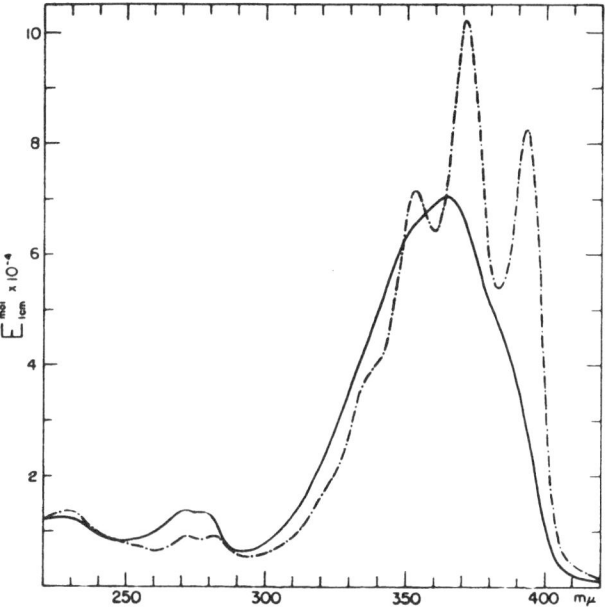

Fig. 88. Molecular extinction curves of *cis*-II-diphenyloctatetraene in hexane: ————, fresh solution, and •—•—•—•, mixture of stereoisomers after 15 min. iodine catalysis (*522*). [From: J. Amer. Chem. Soc. **76**; 4144 (1954).]

Table 33. Some Spectral Characteristics of Stereoisomeric
Diphenyloctatetraenes (in hexane).

Configuration	Position of λ_{max} (mμ)	$E_{1\ cm.}^{mol.} \times 10^{-4}$ at λ_{max}	Fine structure in main band	λ_{max} shift (mμ)	Position of maximum in the *cis*-peak region (mμ)	$E_{1\ cm.}^{mol.} \times 10^{-}$ at *cis*-peak
All-*trans*	371–372	10.8	sharp	0	—	0
Middle-*cis*						
(*cis*-I)	369–370	7.65	moderate	2	273–274, 283	2.8
1-*cis* (*cis*-II)	364	7.0	slight	7–8	271–272, 280	1.3
1,3- or 1,5-di*cis*						
(*cis*-III)	359–360	6.7	none	12	see Fig. 89	(slight)

light-sensitive (Table 32). The stability of *cis*-I is comparable with that of the all-*trans* compound, but *cis*-II and -III are markedly less thermo- and photostable. Concentrated *cis*-II solutions can be kept at 4°, in darkness, for several days without rearrangement, while *cis*-III undergoes considerable spatial change. When stored in the absence of solvents, the oily *cis*-III isomer deposits all-*trans* crystals even in darkness.

The sharply differentiated spectra show the following characteristics: *cis*-I, fine structure in the main band, *cis*-peak present; *cis*-II, no fine structure in the main band, *cis*-peak present; and *cis*-III, no fine structure, no *cis*-peak *(Table 33*, p. 177; *Figs. 87–89)*.

The following configurations have been assigned to these isomers.

Because of the fine structure in the fundamental band a terminal (hindered) *cis* double bond cannot be present in the *cis*-I molecule. This isomer could a priori represent either the middle-mono*cis* (3-*cis*, Model B,

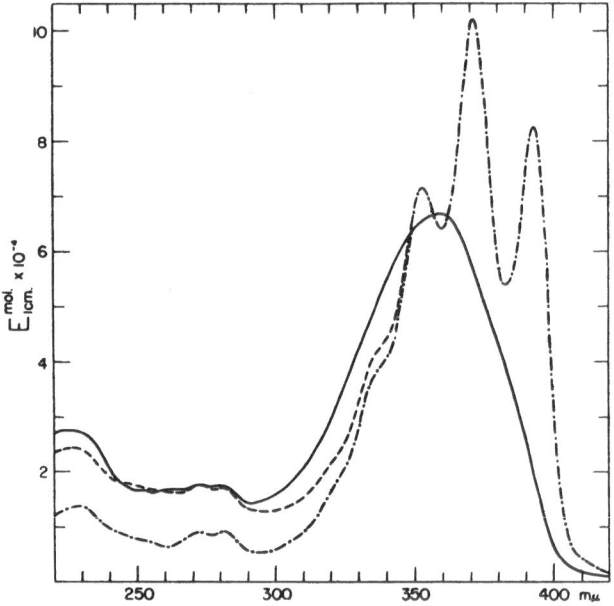

Fig. 89. Molecular extinction curves of *cis*-III-diphenyloctatetraene in hexane: ————, fresh solution; —— ——, mixture of stereoisomers after 15 min. iodine catalysis; and · —· —·, mixture of stereoisomers after catalysis of all-*trans* (minor impurities may be responsible for the higher extinction values of the two *cis*-III curves in the far ultraviolet) *(522)*. [From: J. Amer. Chem. Soc. **76**, 4144 (1954).]

Fig. 86) or the middle di*cis* (3,5-di*cis*) form F. We propose for *cis*-I the 3-mono*cis* configuration considering the very small λ_{max} shift (1–2 mμ) and the high intensity of the *cis*-peak that would contradict the straight overall molecular form of the di*cis* compound mentioned.

The *cis*-II curve (Fig. 88) represents a typical degraded spectrum—a sign of steric hindrance. The presence of a *cis*-peak excludes Models D and E (1,3- and 1,5-di*cis*) which have straight molecular shapes. Hence, we have assigned the 1-*cis* configuration (Model C) to *cis*-II-diphenyl-octatetraene.

Finally, considering the degraded main band and the absence of a *cis*-peak (Fig. 89), a hindered-unhindered di*cis* configuration is proposed for *cis*-III, whereby one is unable to give preference a priori to either the 1,3- or the 1,5-di*cis* configuration (Models D and E). It was pointed out above that such considerations may be supplemented by a study of the rates of iodine-catalyzed rearrangements, because hindered *cis* double bonds rearrange much more rapidly than unhindered ones. As demonstrated in *Figs. 90* and *91*, similar experiments have been useful in the diphenyloctatetraene set. Fig. 90 was obtained by readings at λ_{max} in the fundamental band, and Fig. 91 by those in the *cis*-peak

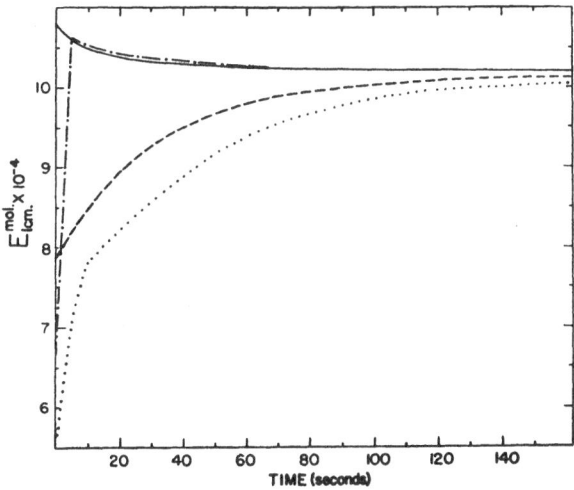

Fig. 90. Changes of molecular extinction values at 371 mμ (in hexane) during iodine catalysis of some stereoisomeric diphenyloctatetraenes: ————, all-*trans*; — — —, *cis*-I; · — · — ·, *cis*-II; and . . . . . ., *cis*-III (*522*). [From: J. Amer. Chem. Soc. **76**, 4144 (1954).]

region. All curves given in these Figures approach finally that of the equilibrium mixture ($\sim$ 86% all-*trans* + 14% *cis* forms).

On iodine catalysis of all-*trans*-diphenyloctatetraene the extinction values decrease slowly at λ_{max} but increase at *cis*-peak. When 3-*cis* (*cis*-I) is catalyzed, the extinction rises slowly in the main band and decreases at about the same rate in the *cis*-peak region, while the molecule rearranges to give mainly the all-*trans* form (Fig. 90). In contrast, the extinction of the main band of the (hindered) 1-*cis* isomer (*cis*-II) at first rises rapidly and then decreases slowly (Fig. 90). Simultaneously, the extinction at *cis*-peak decreases first and then slowly increases (Fig. 91) indicating a rapid conversion to the all-*trans* form. From that point on, the curve almost coincides with the equilibrium curve obtained by the interaction of the all-*trans* form and iodine.

The main-band extinction of 1,3 (or 1,5-)-di*cis*-diphenyloctatetraene (*cis*-III) increases in two distinct steps, the first being rapid and the second slow (Fig. 90); at *cis*-peak the extinction first increases at high rate, reaches a sharp maximum and then decreases (Fig. 91). This behavior is explained by the rapid conversion, di*cis* → 3-*cis*; finally, the latter isomer yields the usual equilibrium mixture.

Evidently, these phenomena confirm the proposed configurations. They also show great similarity in the behavior of diphenylhexatriene and diphenyloctatetraene: On rearrangement with iodine either all-*trans* polyene produces three preferred spatial forms, viz. an unhindered

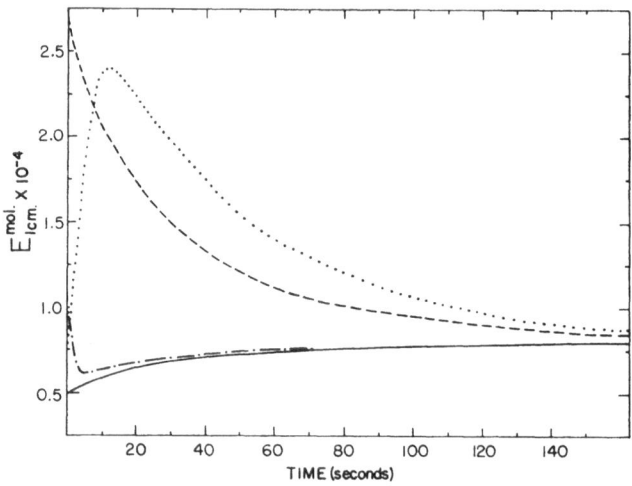

Fig. 91. Progressing changes in the height of the *cis*-peak upon iodine catalysis (measured at 283 mμ, in hexane): ———, starting from all-*trans*-diphenyloctatetraene; — — —, from *cis*-I; · — · — ·, from *cis*-II; and ......, from *cis*-III (522). [From: J. Amer. Chem. Soc. 76, 4144 (1954).]

(middle) mono*cis*, a hindered (peripheral) mono*cis*, and a minor, mono-hindered di*cis* isomer.

Our interpretation of the *cis*-I- and *cis*-II-diphenyloctatetraenes have received a welcome confirmation by a recent study via synthesis. AKHTAR, RICHARDS and WEEDON (4) have prepared the two all-*trans*-diphenyloctatrienyenes, $C_6H_5 \cdot CH=CH—C\equiv C—CH=CH—CH=CH \cdot C_6H_5$ and $C_6H_5 \cdot C\equiv C—CH=CH—CH=CH—CH=CH \cdot C_6H_5$, and obtained by partial hydrogenation over the Lindlar catalyst (p. 62) 3-*cis*- and 1-*cis*-diphenyloctatetraene, respectively. The synthetic preparations were found to be identical with our *cis*-I and *cis*-II samples.

A similar treatment of the diacetylenic compound, $C_6H_5 \cdot CH=CH—C\equiv C—C\equiv C—CH=CH \cdot C_6H_5$, has resulted in the isolation of a new crystalline isomer, to wit the middle-di*cis* (3,5-di*cis*) compound (Model F, Fig. 86) that is clearly different from *cis*-III. Its m. p., 191–193° (then 222–226°) is the highest among the known *cis* members of the set. The spectral curve has fine structure in the fundamental band, although the extinction values are lower than those of the 3-*cis* compound; as predicted (*522*) it does not show a *cis*-peak.

AKHTAR et al. have also reported that the usual 7.03–7.09 μ band (cf. Table 29, p. 160) has appeared in case of the 3,5-di*cis* compound at 7.28 μ, i. e. near the corresponding band of di*cis*-diphenylbutadiene (7.31 μ). If such a shift could be recognized as a general characteristic of two adjacent *cis* double bonds in diphenylpolyenes, then the *cis*-III molecule (band at 7.00 μ) could not include such a diene system and should be assigned the 1,5-di*cis* rather than the 1,3-di*cis* configuration.

The above study of the British authors reaffirms the correctness of the *cis*-peak theory.

For the following diphenylpolyene sets, containing side-chains, which have been synthesized by KARRER's research group, see Chapter IX: 1,18-diphenyl-3,7,12,16-tetramethyl-octadecanonaene (p. 104), 1,3,7,12,16,18-hexaphenyl-octadecanonaene, and 1,18-di-β-naphthyl-3,7,12,16-tetramethyl-octadecanonaene (p. 105); YAMAGUCHI's renieratene and isorenieratene, p. 105.

6. Sets Containing Bulky End Groups and a Short Conjugated System.

2,2′-Diamino-bis-diphenylene-ethylene Set.

KUHN, ZAHN and SCHOLLER (*274*) found that their synthetic sample was a mixture of the *cis* and *trans* isomers. The two forms, Base *A* and Base *B*, were separated chromatographically and obtained as crystals. Interconversion of these isomers takes place on heating, illumination,

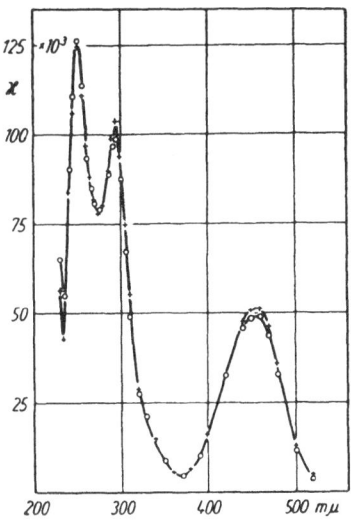

Fig. 92. Molecular extinction curves of *cis-trans* isomeric 2,2′-diamino-bis-diphenylene-ethylene (in 94% alcohol): +++, form *A*; and o o o o, form *B*; according to KUHN, ZAHN and SCHOLLER (*274*). [From: Liebigs Ann. Chem. 582, 196 (1953).]

insolation, or iodine catalysis; a powerful catalyst is colloidal sulfur in benzene. Configurational assignments have not yet been possible, since the spectra are identical both in the UV *(Fig. 92)* and in the IR regions. A and B have been differentiated, however, by the faculty of retaining stoichiometric amounts of certain solvents and by Debye-Scherrer diagrams of such complexes.

(LXIX.) 2,2'-Diamino-bis-biphenylene-ethylenes.

1,2-Bis-[m-nitrophenyl]-1,2-diphenyl-ethylene and 1,2-Bis-[p-nitrophenyl]-1,2-diphenyl-ethylene Sets.

KUHN and BLUM *(259)* should be credited with the resolution, by fractional crystallization, of the compounds (LXX) and (LXXI) into their respective stereoisomers *A* and *B* *(Table 34)*.

(LXX.) 1,2-Bis-[m-nitrophenyl]-1,2-diphenyl-ethylenes.

(LXXI.) 1,2-Bis-[p-nitrophenyl]-1,2-diphenyl-ethylenes.

The UV-curves show near-identity for *m*-nitro *A* and *B*, and similarity for the two corresponding *p*-nitro pair *(Fig. 93)*. Some characteristic differences appear in the IR region: Thus, the *m*-nitro isomer *A* has bands at 11.95 μ and 13.85 μ that are absent from the *B* curve; hence 100% steric purity is postulated for the best *B* samples. Furthermore,

Table 34. Some Characteristics of the Stereoisomeric Forms of
1,2-Bis-[*m*-nitrophenyl]-1,2-diphenyl-ethylene and the Corresponding
p-Nitro Compound (*259*).

Compound	Steric form	Steric purity of sample (%)	Crystal form	Color of crystals	M. p.	Soluble in ethanol	Maxima in ethanol (mμ)
m-NO$_2$	A	> 90	prisms	yellow	156–157°	more	244 (310)
	B	100	needles	light yellow	188–189°	less	236 (310)
p-NO$_2$	A	> 90	leaflets	dark yellow	215–216°	more	267, 365
	B	100	small prisms	yellow	228–229°	less	256, 330

some intense bands present in the *B* curve scarcely appear in case of *A*;
the best *A* sample is supposed to contain less than 10% *B*.

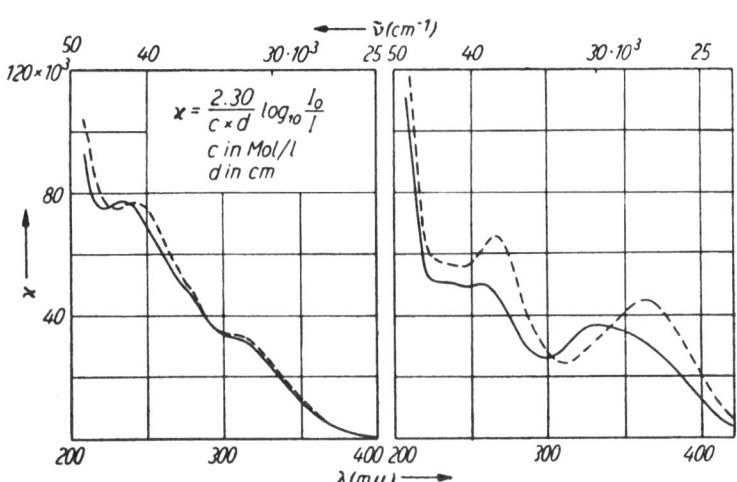

Fig. 93. Left: Extinction curves of *cis-trans* isomeric 1,2-bis-[*m*-nitrophenyl]-1,2-diphenyl-ethylene, in abs. ethanol: — — —, isomer *A* and ————, isomer *B*. Right: the corresponding curves of the *p*-nitro compound. According to Kuhn and Blum (*259*). [From: Chem. Ber. 92, 1483 (1959).]

Although no final configurational assignments can be made, strong
indications are given by the respective dipole moments. These values
are, for *m*-nitro *A* and *B*, 5.7 D and 4.9 D; for *p*-nitro *A* and *B*, 1.9 D
and 7.0 D. It is assumed that in the *m*-nitro set the *cis* configuration
is more probable for *A* than for *B*; and such assignment would be in
accordance with the observed unilateral thermal rearrangement *B* → *A*.
In the *p*-nitro set the value 7.0 D lies within the range calculated for
the *cis* isomer (m. p. 216°), the value 1.9 D for *trans* (m. p. 229°) is
explained by some contamination with the *cis* form.

1-Biphenylene-4-phenylbutadiene, 1-Biphenylene-6-phenylhexatriene, 1,6-Di-biphenylene-hexatriene, and 1,1,6,6-Tetraphenyl-hexatriene Sets.

The *cis* forms of the first three compounds were described by Magoon and the writer (*299*). As *Figs. 94–96* show, 1-biphenylene-phenylbutadiene is expected to yield a single *cis* isomer that is sterically hindered; 1-biphenylene-6-phenylhexatriene, three *cis* forms, i. e. the unhindered 3-*cis*, the hindered 5-*cis* and the hindered-unhindered 3,5-di*cis*; and di-biphenylene-hexatriene, a single, unhindered *cis* form. All these stereoisomers have been prepared. They fluoresce in UV light and appear

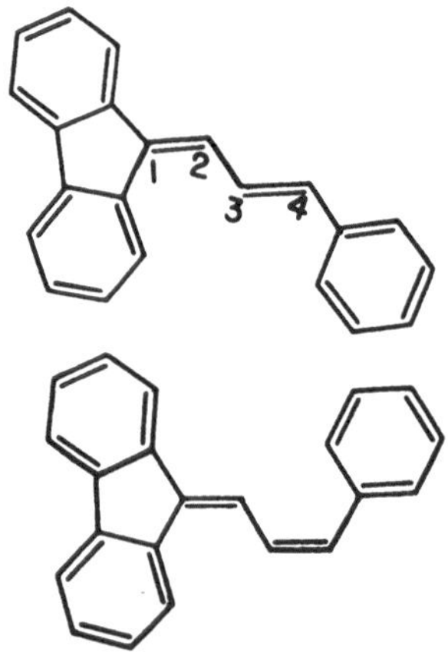

Fig. 94. Models of the two stereoisomeric 1-biphenylene-4-phenylbutadienes: top, *trans*; and bottom, *cis*.

below the corresponding *trans* zone on the column. The weakening of the adsorption affinity is especially marked in the presence of a hindered *cis* double bond.

Since the iodine-catalyzed equilibria contain only 2–8% *cis* form(s), a very brief heating of fused all-*trans* crystals at 320–340° is the preferred preparative procedure; it yields 10–20% *cis* isomers in the recovered substance (loss, 20–25%).

As expected, no *cis*-peaks appear in the spectral curves, because in this type of compound *trans → cis* rearrangements do not markedly alter the straight overall shape of the short aliphatic conjugated system. Hindered *cis* isomers show degraded fundamental bands *(Figs. 97–99)*.

Concerning relative stabilities, the following statements are valid for the three sets: Under the conditions of refluxing, in darkness, the *trans* and *cis* configurations are equally resistant. However, each *cis* form, especially when sterically hindered, is markedly more photosensitive than the corresponding *trans* compound. On iodine catalysis the conversion rates are much higher for hindered than for unhindered *cis* double bonds—a general phenomenon.

1-Biphenylene-4-phenylbutadiene. The *trans* form did not change on irradiation, illumination or refluxing. The oily *cis* form was photo-

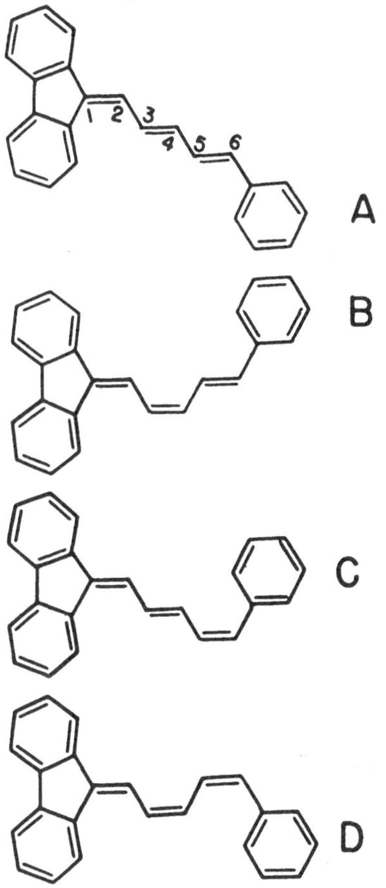

Fig. 95. Models of the four stereoisomeric 1-biphenylene-6-phenylhexatrienes: *A*, all-*trans*; *B*, 3-mono*cis* [(assigned to *cis*-I); *C*, 5-mono*cis*(*cis*-II); and *D*, 3,5-di*cis* (*cis*-III).

sensitive; in the absence of solvents it deposited all-*trans* crystals even in quasi-darkness. The main band was practically void of fine structure *(Fig. 97)* and the λ_{max} shift, 15.5 mμ, almost equalled that observed in the diphenylbutadiene set.

1-Biphenylene-6-phenylhexatriene. Neither irradiation, insolation, nor refluxing affected the all-*trans* configuration but fusion experiments

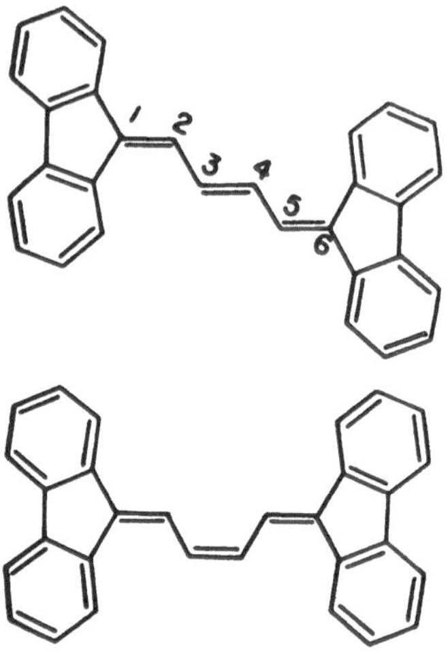

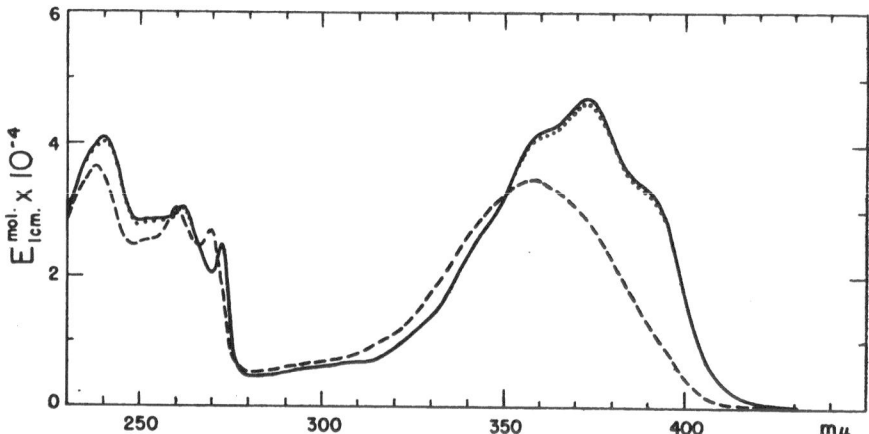

Fig. 96. Models of the two stereoisomeric 1,6-di-biphenylene-hexatrienes: top, *trans*; and bottom, *cis*.

Fig. 97. Molecular extinction curves of 1-biphenylene-4-phenylbutadienes (in hexane): ————, fresh solution of the *trans* form; — — —, fresh solution of the *cis* form; and ......, mixture of stereoisomers after iodine catalysis (the dotted line differs only slightly from the full line) (*299*). [From: J. Amer. Chem. Soc. **77**, 5642 (1955).]

did yield the three expected *cis* forms, all crystallizable and termed I–III, in the order of decreasing adsorbabilities. In the recovered pigment the ratio was, unchanged all-*trans* : *cis*-I : *cis*-II : *cis*-III = 79 : 16 : 4 : 1. The configurations have been established as follows.

The *cis*-I curve showed considerable fine structure and a λ_{max} shift of only 2.5 mμ. This isomer clearly represents the only possible unhindered *cis* member of the set, i. e. the 3-mono*cis* form (Model B, Fig. 95). *Cis*-II and -III are sterically hindered and show degraded spectra *(Fig. 98)*. The fundamental band of *cis*-III is located at shorter wavelengths and has lower extinction values, hence it is interpreted as the 3,5-di*cis* form (Model D) which leaves the 5-*cis*-configuration for *cis*-II (Model C).

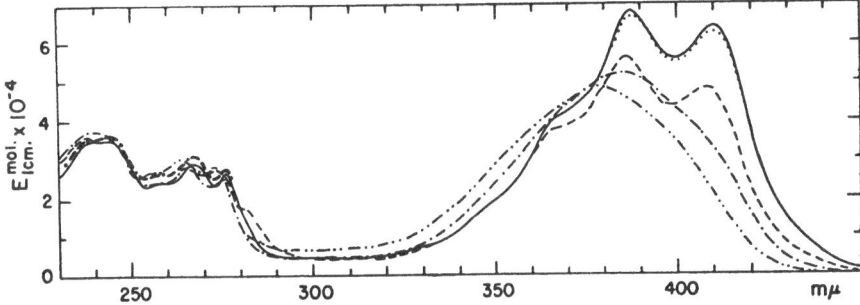

Fig. 98. Molecular extinction curves of 1-biphenylene-6-phenylhexatrienes (in hexane): ————, fresh solution of the all-*trans* form; — — —, fresh solution of 3-mono*cis* (*cis*-I); · —·—·, of 5-mono*cis* (*cis*-II); ··—··—··, of 3,5-di*cis* (*cis*-III); and ......, mixture of stereoisomers after iodine catalysis (*299*). [From: J. Amer. Chem. Soc. 77, 5642 (1955).]

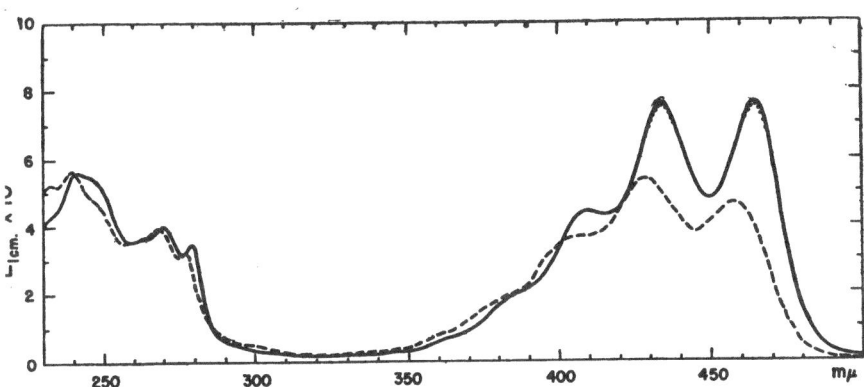

Fig. 99. Molecular extinction curves of 1,6-di-biphenylene-hexatrienes (in cyclohexane): ————, fresh solution of the *trans* form; — — —, fresh solution of the *cis* form; and ......, mixture of stereoisomers after iodine catalysis (the dotted line differs from the full line only slightly at the main maxima) (*299*). [From: J. Amer. Chem. Soc. 77, 5642 (1955).]

These assignments have been confirmed as follows: On iodine catalysis of *cis*-III under very mild conditions the recovered pigment contained 80% *cis*-I, according to the stepwise conversion, 3,5-di*cis* (hindered-unhindered) → 3-mono*cis* (unhindered) → all-*trans*; the hindered 5-mono-*cis* (*cis*-II) was absent and could be obtained only by rearranging the all-*trans* compound.

1,6-Di-biphenylene-hexatriene. As expected, the curve of the unhindered *cis* isomer shows considerable fine structure *(Fig. 99)*; λ_{max} shift, 6.5 mμ.

When refluxed, both forms are equally stable. Illumination of the *trans* isomer with a Photoflood bulb yielded 10% *cis* form; and melting crystals gave 16% *cis* in the recovered pigment. Recently, the *cis* form has been obtained by KUHN and FISCHER (*262*) who conducted a stereospecific reduction of bis-diphenylene-hexapentaene by means of the Lindlar catalyst.

The cumulene, *cis-tetraphenyl-hexatriene* (LXXII) has been prepared in the same manner:

$$R \atop R \bigg\rangle C{=}C{=}C{=}C{=}C{=}C \bigg\langle {R \atop R} \quad \xrightarrow{+2\,H_2} \quad$$

(LXXII.) *cis*-Tetraphenyl-hexatriene $(R = C_6H_5)$.

Such partial reductions of cumulenes constitute the final step in this new, promising method of *cis*-polyene synthesis, because in some instances it is easier to build up a cumulene than the corresponding conjugated polyene.

XV. Cumulenes with Aromatic Terminal Groups.

As is well known, the cumulenes (in German, "Kumulene") possess n double bonds attached to $n + 1$ carbon atoms, the simplest representatives being allenes (propadienes) $C=C=C$ and butatrienes $C=C=C=C$. They can be converted into *cis* polyenes (pp. 58, 62, 188).

In his classical treatise, VAN'T HOFF (*444, 445*) predicted as early as 1875 the existence of geometrical isomerism around such systems:

"... ma théorie prévoit deux isomères, quand deux atomes de carbone sont doublement liés, soit directement, soit par intermédiaire de plusieurs carbones à double liaison."

Although some allenes have been resolved into their optical antipodes (*251, 302, 303*), it took about 80 years until VANT' HOFF's prediction could be realized experimentally (*258 a*). The first attempts to obtain *cis* and *trans* forms remained unsuccessful in the case of some tetraaryl-butatrienes (*264*); likewise, only a single distyryl-butatriene (LXXIII) could be obtained. Interestingly, its spectrum showed a marked *cis*-peak at 280 mμ, i. e. at a distance of 150 mμ from the longest wavelength maximum. The sample probably contained one or more *cis* forms [KUHN and KRAUCH (*265*)].

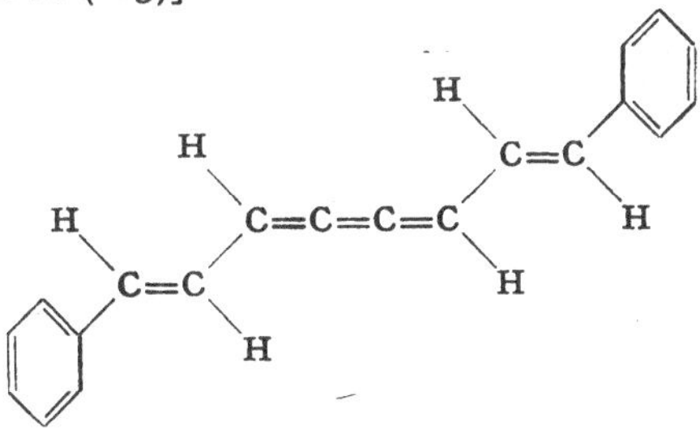

(LXXIII.) 1,4-Distyrylbutatriene (all-*trans*).

A decisive advance in this field was initiated by KUHN and SCHOLLER (*267*) who recognized that a postulate for the chromatographic resolvability of aromatic *cis-trans* cumulene mixtures is the presence of certain substituents such as NO_2, NH_2, or $NHCOCH_3$. Thus, successful resolutions were carried out in the bis-[2-nitrodiphenylene]-butatriene (LXXIV) and 1,4-bis-[*m*-nitrophenyl]-1,4-diphenyl-butatriene (LXXV) sets.

The UV spectra of the respective *cis-trans* pairs are practically identical, but the IR spectra and the Debye-Scherrer diagrams do show marked differences. Because no unambiguous configurational assignments have been possible in these sets, the individual isomers were termed *A* and *B*, the symbol *A* being used for that isomer whose zone occupied the lower position on the column. The best *A* sample contained about 7% *B* and the best *B* sample 16% *A*; both were crystalline.

Bis-[2-nitrodiphenylene]-butatriene Set.

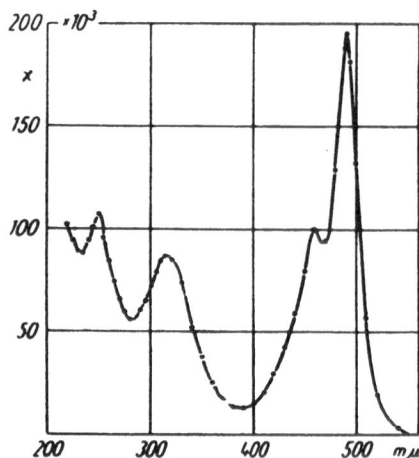

(LXXIV.) Bis-[2-nitrodiphenylene]-butatrienes.

In order to estimate the ratio *A* : *B* the solution was shaken with an α-bromo-naphthalene-bromobenzene mixture and filtered onto an alumina column. After eluting zone *A* with bromobenzene and *B* with bromobenzene-benzonitrile (9 : 1) the components were determined photometrically.

The chromatographic experiments must be carried out in darkness. When during the developing process the column in exposed to the light of a north window, the two stereoisomeric zones flow together. But they separate again upon turning the column by 180°.

Fig. 100. Extinction curve of bis-[2-nitrodiphenylene]-butatrienes *A* and *B*, in dioxane; according to KUHN and SCHOLLER (267). [From: Chem. Ber. 87, 598 (1954).]

As already noted, the two nitrotrienes cannot be differentiated by their UV spectra (λ_{max} at 491 mμ in dioxane; *Fig. 100*). Some regions of the IR curves are, however, so different that they offer a basis for the estimation of the $A : B$ ratio in mixtures. Thus, the bands present in the A curve at 9.4 μ and 12.7 μ have all but disappeared in the best B preparations [OTTING (*342*)]. For Debye-Scherrer diagrams see *Fig. 101*.

The spontaneous stereoisomerization of the two forms in bromo-benzene solution yields the equilibrium, 57% $A \rightleftarrows 43\%$ B.

Fig. 101. Debye-Scherrer diagrams of bis-[2-nitrodiphenylene]-butatrienes: A (top) and B (bottom); according to KUHN and BLUM (*259*). [From: Chem. Ber. **92**, 1483 (1959).]

The *trans* configuration is much more probable for isomer A. The arguments for and against such interpretation have been discussed by KUHN and SCHOLLER (*267*) as well as by OTTING (*342*); see also (*258a*).

Dipole moments that have allowed unequivocal configurational assignments in the case of azobenzene cannot be used in this set, considering the near-insolubility of the substance in non-polar solvents and the limited stability of the solutions. Possibly, nuclear magnetic resonance studies would help.

1,4-Bis-[m-nitrophenyl]-1,4-diphenyl-butatriene Set.

(*cis*) O_2N C=C=C=C NO_2

(*trans*) O_2N C=C=C=C NO_2

(LXXV.) 1,4-Bis-[*m*-nitrophenyl]-1,4-diphenyl-butatrienes.

In contrast to the corresponding *p*-compound, this *m*-substituted butatriene can be resolved into two crystalline isomers, A and B, chromatographically or by heating the synthetic *cis-trans* mixture to 160°, pressed in a KBr pellet (*259*). Under the conditions applied the thermic equilibrium was practically on the side of A which was obtained 100% pure (cf. the analogous behavior of dimethylcrocetin, *263*). The best B preparations were only 80% pure.

A difficulty in handling either isomer is caused by photosensitivity: even weak diffuse daylight rapidly yields an $A \rightleftarrows B$ equilibrium. When after successful chromatographic separation in darkness, one side of the column was turned towards a window, while the developing process was still going on, the two zones were washed together on the illuminated side.

The UV spectra of A and B are almost identical (λ_{max} at 426 mμ, in benzene; *Fig. 102*), but marked differences exist in the IR region. Conspicuous is the absence of the 15.10 μ band in the A curve. *Fig. 103* shows the change of the IR spectrum upon heating ($B \rightarrow A$).

The dipole moments are, 5.2 D for A and 5.9 D for B. These values would favor the *cis* configuration for B and *trans* for A, in accordance with the higher thermostability of A.

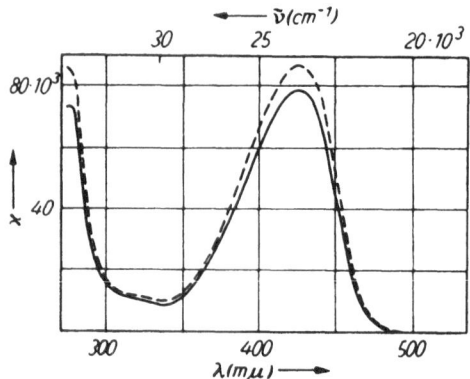

Fig. 102. Spectral curves of *cis-trans* isomeric 1,4-bis-[*m*-nitrophenyl]-1,4-diphenylbutatrienes, in benzene: — — —, isomer A; and ———, isomer B; according to Kuhn and Blum (*259*). [From: Chem. Ber. 92, 1483 (1959).]

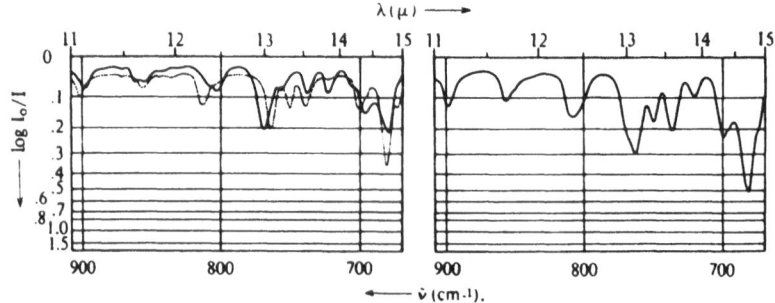

Fig. 103. Infrared curves of *cis-trans* isomeric 1,4-bis-[*m*-nitrophenyl]-1,4-diphenylbutatrienes, in KBr. Left: Isomer B, ———, before; and — — —, after heating (rearrangement to A). Right: After subsequent illumination (mixture of A and B); according to Kuhn and Blum (*259*). [From: Chem. Ber. 92, 1483 (1959).]

XVI. Polyene Azines.

$$C_6H_5 \cdot (CH=CH)_n \cdot CH=N—N=CH(CH=CH)_n \cdot C_6H_5$$

It was interesting to extend the study of diarylpolyenes to this symmetrical series that is characterized by a centrally located $=N—N=$ group. Although the lowest homolog, benzalazine ($n = 0$) has given negative results, cinnamalazine ($n = 1$) and phenylpentadienalazine ($n = 2$) could be converted by DALE and the writer (*84*) into two oily stereoisomers each, viz. a *cis* $C=C$ and a *cis* $N=C$ form. One of the *cis* cinnamalazines is sterically hindered.

Cinnamalazine Set.

The calculated number of steric forms in this set is 10, among them 2 mono*cis* compounds. Only two *cis* forms "I" and "II" have been

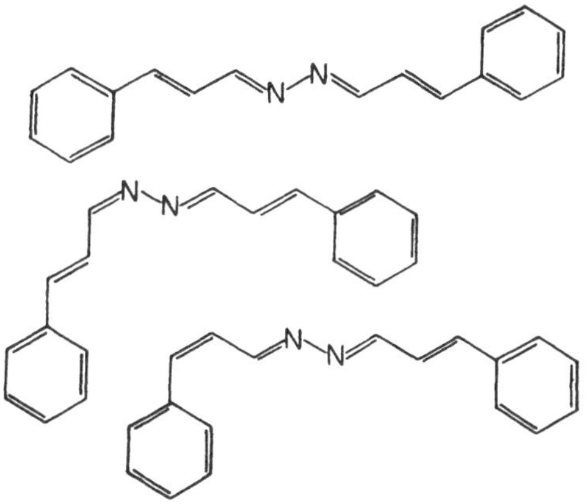

Fig. 104. Models of three spatial forms of cinnamalazine: top, all-*trans*; middle, proposed configuration for *cis*-I (*3-cis*); and bottom, for *cis*-II (*1-cis*) (*84*). [From: J. Amer. Chem. Soc. 75, 2379 (1953).]

observed ($\lambda_{\max}$ shifts, 2–3 mμ) *(Fig. 104)*. On the column *cis*-I was located above but *cis*-II just below the all-*trans* zone; both *cis* zones appeared dull-red in ultraviolet light. When the oily *cis*-I isomer was kept in darkness at 4° for several days or exposed to sunshine for a few minutes, crystals of newly formed all-*trans* azine were deposited.

The following treatments converted all-*trans*-cinnamalazine partially into its *cis*-I form: Standing in solution at room temperature, overnight;

Table 35. Composition of Stereoisomeric Mixtures of Cinnamalazines
Obtained from the All-*trans* Form.

Treatment	Content in the recovered substance (%)			Recovery (%)
	unchanged all-*trans*	*cis*-I	*cis*-II	
Sunshine, 1 hr.	75	23	1–2	
Sunshine, 6 hr.	68	26	6	
Sunshine, iodine, 1 hr.	74	23	3	
Sunshine, iodine, 6 hr.	67	25	8	
Daylight lamp, 1 hr.	80	18	1–2	~ 100
Daylight lamp, iodine, 1 hr.	81	17	1–2	
Daylight lamp, iodine, 6 hr.	82	17	1–2	
Refluxing in hexane, 3 hr.	78	20	1–2	
Melting crystals, 175°, 2 min.	73	23	4	80
Melting crystals, 190°, 10 min	70	21	8	25

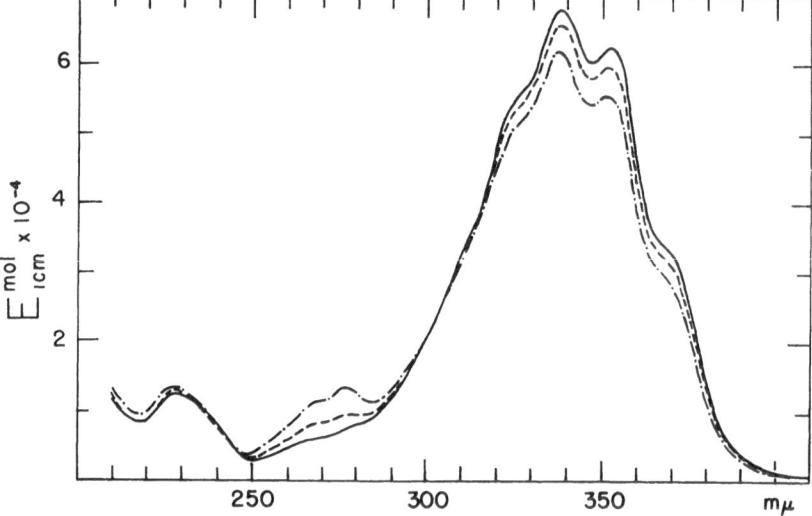

Fig. 105. Molecular extinction curves of cinnamalazine in hexane: ———, fresh solution of the all-*trans* form; — — —, mixture of stereoisomers upon exposure to daylight lamp for 150 min. (no equilibrium was reached at this point; the same curve was obtained in the presence of iodine; approximately the same curve was obtained upon refluxing in dim light for 6 hr.); and · — · — ·, mixture of stereoisomers after 10 min. insolation (constant after 30 min.) *(84)*. [From: J. Amer. Chem. Soc. **75**, 2379 (1953).]

refluxing in dim light for an hour; exposure to a daylight lamp or sunshine. When all-*trans* crystals were melted or dilute solutions were subjected to prolonged insolation, *cis*-II also appeared. As compared to most carotenoids the all-*trans* form suffered only relatively slight spectral change when catalyzed by iodine, in light, because of the predominance of the all-*trans* form in the equilibrium *(Table 35)*.

The two *cis* forms represent different types. A hexane solution of *cis*-II remains practically unchanged during 2 hours refluxing but *cis*-I rearranges to a large extent at 20° within a few hours. Moreover, *cis*-I is by far more light-sensitive than *cis*-II. The relative photosensitivities are, however, inverted in the presence of iodine, and then the difference in the behavior of the two *cis* forms becomes still more striking: The moderate photo-rearrangement rates of *cis*-I are not markedly increased

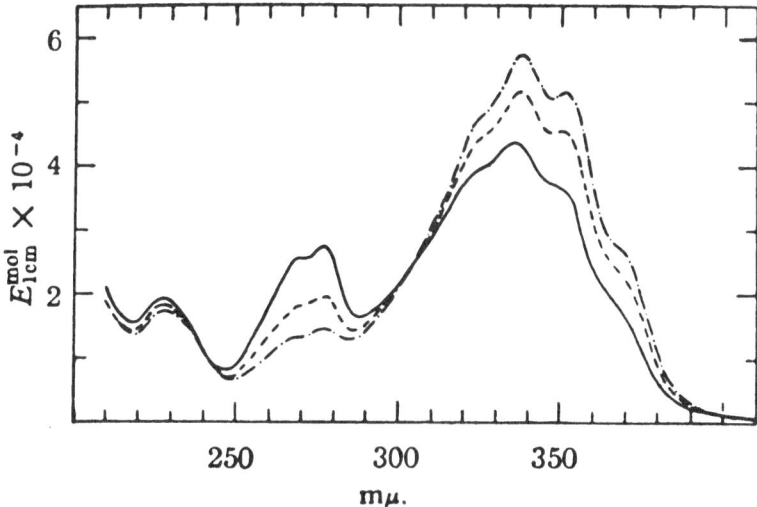

Fig. 106. Molecular extinction curves of *cis*-I-cinnamalazine in hexane: ————, fresh solution; — — —, mixture of stereoisomers after exposure to daylight lamp for 150 min. (no equilibrium was reached at this point; the same curve was obtained in the presence of iodine); and · — · — ·, mixture of stereoisomers after 10 min. insolation of *cis*-I (constant after 15 min.) (*84*). [From: J. Amer. Chem. Soc. **75**, 2379 (1953).]

by the catalyst that has a powerful effect on the otherwise rather photo-stable *cis*-II. With some oversimplification *cis*-I may be designed as iodine-resistant and *cis*-II as iodine-sensitive.

Fig. 105 illustrates the spectroscopic behavior of cinnamalazine. The isomer mainly responsible for the *cis*-peak effect (at 277 mμ in hexane) is photochemically formed *cis*-I-cinnamalazine whose spectral curve shows a very high *cis*-peak *(Fig. 106)*; on illumination (especially insolation) the *cis*-peak of *cis*-I decreases rapidly while the height of the main band increases. Much smaller is the spectral shift when all-*trans*-cinnamalazine is refluxed, indicating that the thermal equilibrium mixture contains more of the all-*trans* form than results from illumination. Accordingly, when *cis*-I is refluxed, the *cis*-peak flattens and the formation of all-*trans* azine is more complete than in photochemical experiments (*Fig. 107*, next page).

The *cis*-II-cinnamalazine curve *(Fig. 108)* differs conspicuously from those of the *trans-* and *cis-*I forms by the absence of fine structure in both the fundamental and *cis*-peak regions. In a way, the pair, all-*trans-* and *cis*-II-cinnamalazine, parallels the pair, all-*trans-* and mono*cis*-diphenylbutadiene.

During brief exposure of *cis*-II to a daylight lamp, in the presence of iodine, the curve flattened out in the *cis*-peak region, the main maximum

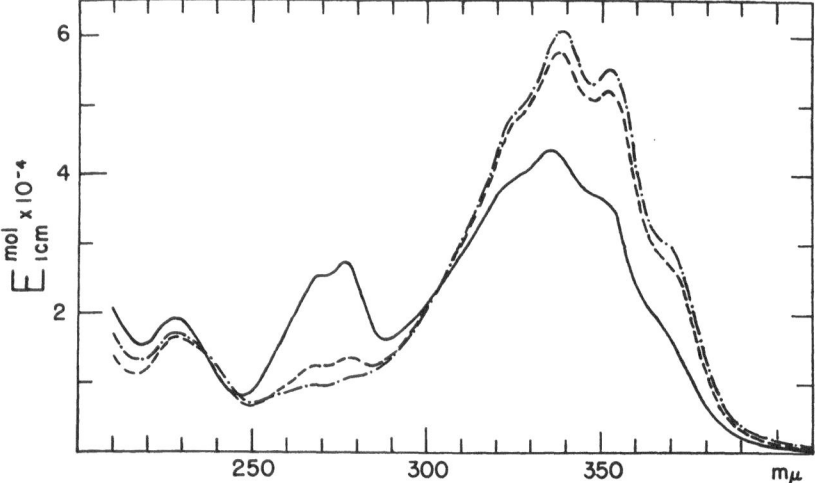

Fig. 107. Molecular extinction curves of *cis*-I-cinnamalazine in hexane: ————, fresh solution; — — —, mixture of stereoisomers after refluxing in dim light for 60 min.; and · —·—·, after 120 min. (practically constant after 180 min.) *(84)*. [From: J. Amer. Chem. Soc. 75, 2379 (1953).]

increased and considerable fine structure appeared, which demonstrates the almost exclusive formation of the all-*trans* compound (Fig. 108). On prolonged illumination, however, the *cis*-peak developed anew and the main band somewhat decreased, because marked amounts of *cis*-I were formed, originating from a rearrangement of the all-*trans* isomer formed during the first phase of the experiment.

The interpretation of these spectroscopic features has been confirmed by chromatographic separations.

In accordance with its high *cis*-peak we assign to *cis*-I-cinnamalazine a *cis* configuration at a C=N bond, i. e. at the nearest possible position to the center (Fig. 104, middle, p. 193). Evidently, the behavior of *cis*-I in iodine catalysis, which is quite different from that of *cis* C=C bonds in carotenoids or diphenylpolyenes, indicates a special steric feature. The small λ_{max} shift (2–3 mμ) also points to a mono*cis* and excludes a tri*cis* configuration. Neither can *cis*-I be a di*cis* compound, because as models show, two of the four possible di*cis* cinnamalazines can display only very low *cis*-peaks and the two others none at all.

The configuration proposed for *cis*-II (Fig. 104, bottom) represents the second possible mono*cis* form, in accordance with the moderate *cis*-peak. The lack of fine structure and the iodine sensitivity are also explained by this hindered configuration. A di*cis* configuration is excluded

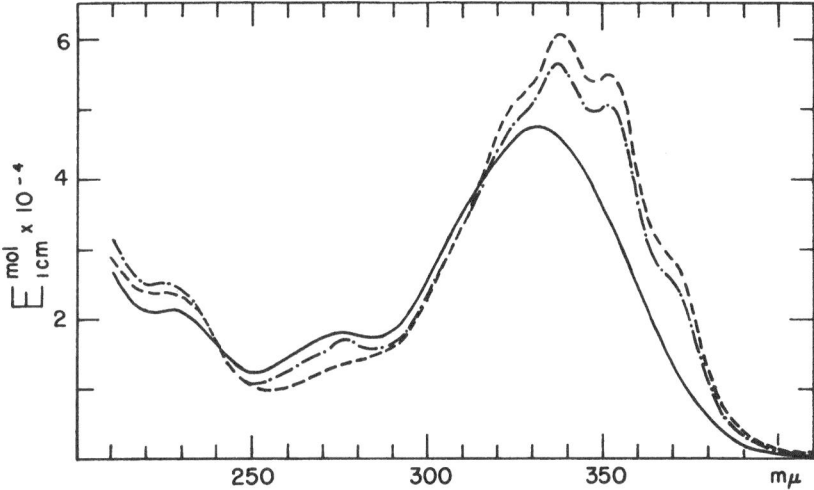

Fig. 108. Molecular extinction curves of *cis*-II-cinnamalazine in hexane: ————, fresh solution (this curve remained unchanged during illumination with a daylight lamp for 150 min.); — — —, mixture of stereo-isomers after 5 min. exposure to daylight lamp in the presence of iodine (exposure for 150 min. caused only a slight decrease in the main band and a slight increase in the *cis*-peak region); and · —·—· mixture of stereoisomers after 5 min. insolation of either of the solutions illuminated with daylight lamp or of a fresh solution, in the presence of iodine. (In the absence of the catalyst more than 1 hr. insolation was required for obtaining a curve of approximately the same shape; then a general decrease in the extinction indicated partial destruction.) *(84)*. [From: J. Amer. Chem. Soc. **75**, 2379 (1953).]

because of the considerable *cis*-peak; neither is the tri*cis* configuration suggested because *cis*-II yields on illumination first the all-*trans* and only later (in part) the *cis*-I form. It would be unclear why tri*cis*-cinnamalazine should not afford *cis*-I directly, rather than via the all-*trans* isomer.

Phenylpentadienalazine Set.

When this azine *(Fig. 109)* was submitted to the treatments mentioned in *Table 36*, it afforded in each instance two isomers, *cis*-I and *cis*-II, the former being the main product [DALE and the writer *(84)*]. Catalysis by iodine under a daylight lamp rearranged a much smaller fraction of the all-*trans* form than in the case of carotenoids. The presence of iodine markedly shifted the equilibrium in favor of *cis*-II-phenylpentadienalazine. Thus, after 3 hours treatment the ratio, *cis*-I : *cis*-II was 9 : 1 in the absence, but 3 : 1 in the presence of the catalyst.

When developed on activated zinc carbonate with hexane containing 5–8% acetone, the *cis*-I zone appeared much above the all-*trans* form. This sequence is, however, inverted on weak zinc carbonate. On the same adsorbent of medium activity no separation takes place. *cis*-II-Phenylpentadienalazine separates from the all-*trans* compound easily and always appears below it.

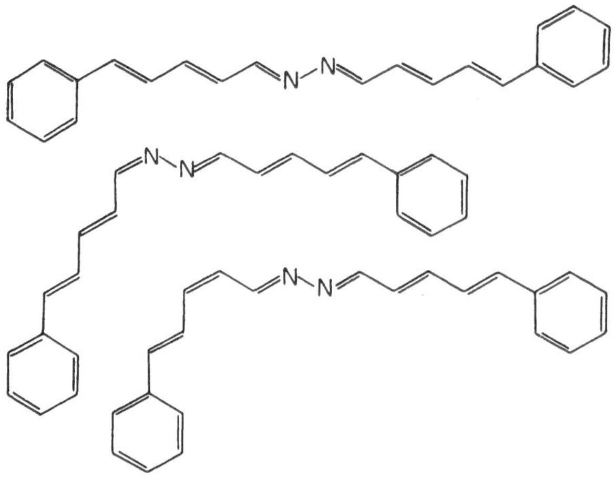

Fig. 109. Models of three steric forms of phenylpentadienalazine: top, all-*trans*; middle, probable configuration of *cis*-I (5-*cis*); and bottom, of *cis*-II (3-*cis*) (*84*). [From: J. Amer. Chem. Soc. **75**, 2379 (1953).]

The relative stabilities differ under various conditions. Like *cis*-I-cinnamalazine, the oily *cis*-I-phenylpentadienalazine rearranges in the absence of solvents and spontaneously yields all-*trans* crystals. This process was found to be practically complete within 1–2 weeks in the

Table 36. Composition of Stereoisomeric Mixtures of Phenylpentadienalazines Obtained from the All-*trans* Form.

Treatment	Content in the recovered substance (%)			Recovery (%)
	unchanged all-*trans*	*cis*-I	*cis*-II	
Sunshine, 1 hr.	67	24	9	92
Sunshine, iodine, 1 hr.	70	21	9	91
Daylight lamp, 1 hr.	85	13	2	92
Daylight lamp, 3 hr.	80	18	2	92
Daylight lamp, iodine, 1 hr.	83	12	5	93
Daylight lamp, iodine, 3 hr.	77	17	6	91
Refluxing in hexane, 3 hr.	78	20	2	82
Melt in naphthalene, 160°, 2 min. . .	76	21	3	70
Melt in naphthalene, 160°, 10 min. .	74	22	4	60

cold room, in darkness, but it took place almost instantaneously in bright sunshine. No crystals appeared when *cis*-II was kept in darkness; and in sunshine several minutes were required for crystallization.

When a hexane solution was refluxed in dim light for 1–2 hours, *cis*-I-phenylpentadienalazine underwent extensive rearrangement but

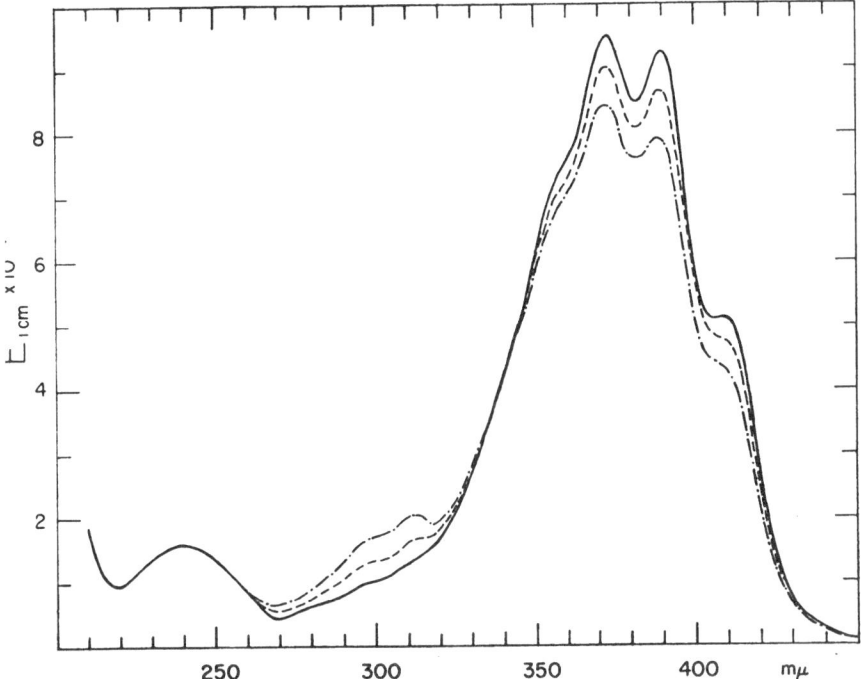

Fig. 110. Molecular extinction curves of phenylpentadienalazine in hexane: ————, fresh solution of the all-*trans* form; — — —, mixture of stereoisomers upon 10 min. exposure to daylight lamp; and ·—·—·, after 150 min. The same values were observed in the presence of iodine. A curve very similar to the dashed one was also obtained by 10 min. insolation (*84*). [From: J. Amer. Chem. Soc. 75, 2379 (1953).]

also some irreversible destruction which is the main process in the case of *cis*-II.

In sunshine, solutions of all three steric forms rearrange rapidly and yield the same mixture; no significant destruction takes place. More sharply dependent on the configuration is the rearrangement induced by illumination with a daylight lamp; this exposure of either the all-*trans* or the *cis*-I form caused marked stereoisomerization while *cis*-II remained almost unaltered even during 3 hours. Under the influence of iodine, however, most of *cis*-II was converted into the *trans* form within 5 minutes; the two other isomers were much less affected (Table 36).

While illuminating all-*trans*-phenylpentadienalazine with a daylight
lamp, the main maxima decreased and a *cis*-peak grew out at 311 mμ
(in hexane; *Fig. 110*). Refluxing had a less pronounced spectral effect.
Both pure *cis* forms showed a λ_{max} shift of 2–3 mμ and similar *cis*-
peaks.

There is, however, one remarkable difference in their respective
behavior toward iodine: Under the influence of a daylight lamp the

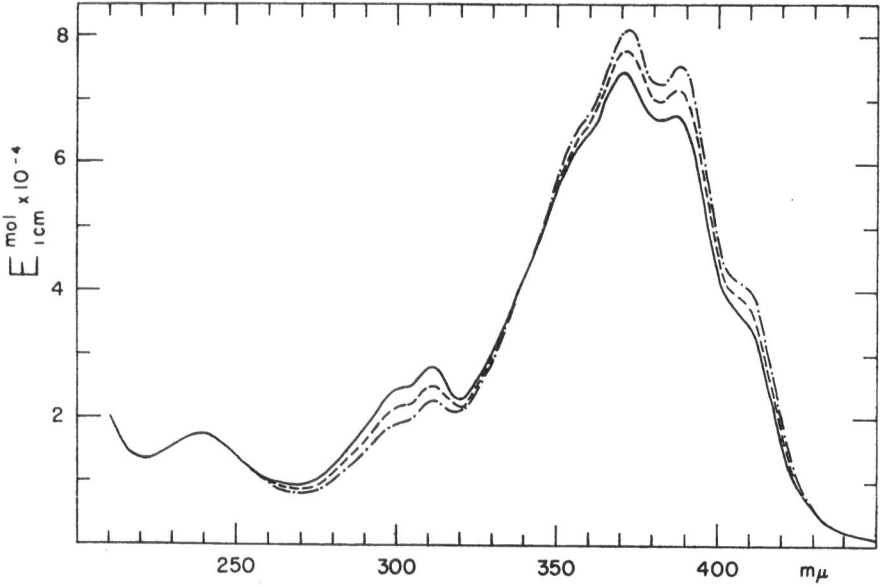

Fig. 111. Molecular extinction curves of *cis*-I-phenylpentadienalazine in hexane: ————, fresh solution;
— — —, after exposure to daylight lamp for 10 min.; and · — · — ·, after 30 min. (constant after 150 min.).
Identical curves were obtained in the presence of iodine. A curve very similar to · — · — · was obtained
upon 10 min. insolation (constant after 30 min.) (84). [From: J. Amer. Chem. Soc. 75, 2379 (1953).]

cis-peak of *cis*-I decreased, the extinction values in the main band
increased within 10 min., and this shift was *not* influenced by iodine
(Fig. 111). A similar exposure had almost no effect on *cis*-II in the course
of 2 hours; a trace of iodine, however, did induce rapid stereoisomerization
and gave within 5 min. a curve that included a low *cis*-peak. As in the
case of *cis*-II-cinnamalazine, this decrease in the *cis*-peak was followed,
upon prolonged illumination, by a distinct increase as a result of the
secondary process, all-*trans* → *cis*-I *(Fig. 112)*.

We submit that *cis*-I-phenylpentadienalazine is an analog of *cis*-I-
cinnamalazine and represents that mono*cis* form in which a C=N bond
has assumed *cis* configuration (Fig. 109, middle). This is in accordance
with the position of the main maxima, the relatively high *cis*-peak and

the described "abnormal" behavior in iodine catalysis. In contrast, *cis*-II must possess a C=C mono*cis* configuration. The considerable height of its *cis*-peak and the degree of fine structure in the main band (similar to that of the corresponding all-*trans* band but very different from the *cis*-II-cinnamalazine band) exclude a peripheral (hindered) location. Hence, the configuration shown in Fig. 109 (bottom) is proposed for *cis*-II-phenylpentadienalazine (p. 198).

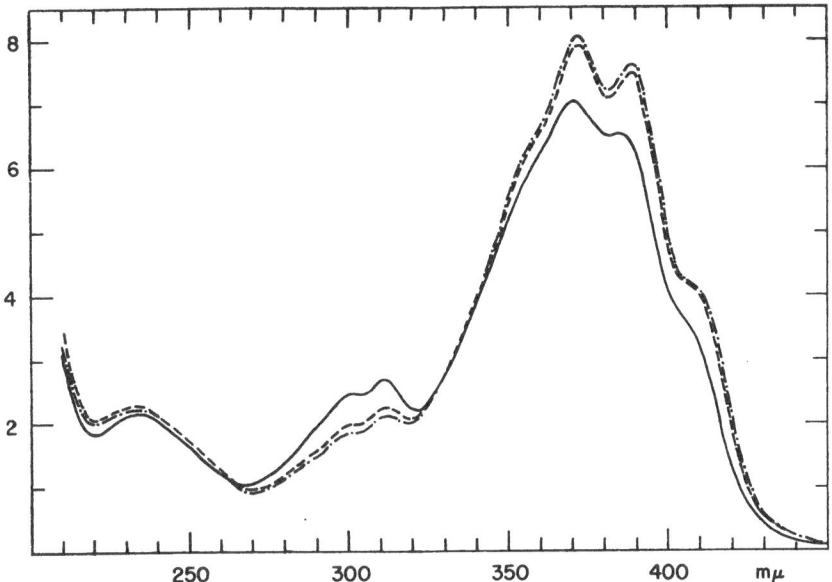

Fig. 112. Molecular extinction curves of *cis*-II-phenylpentadienalazine in hexane: ————, fresh solution (constant during a 30 min. illumination with the daylight lamp in the absence of iodine); — — —, after illumination, in the presence of iodine, for 5 min.; and ·—·—·, for 10 min. After 150 min. the extinction in the λ_{max} region decreased by 5% while in the *cis*-peak region a slight increase (up to the 5-min. curve) was observed (*84*). [From: J. Amer. Chem. Soc. **75**, 2379 (1953).]

XVII. Bibliography.

1. ABDULLAH, M. M., S. R. MORCOS and M. K. SALAH: Content of Vitamin A₂ in some Nile Fishes. Biochemic. J. **56**, 569 (1954).
2. AHMAD, R. and B. C. L. WEEDON: Carotenoids and Related Compounds. I. Total Synthesis of "all-*trans*"-Methylbixin and of a Diketone with the Capsorubin Chromophore. J. Chem. Soc. (London) **1953**, 3286.
3. — — Carotenoids and Related Compounds. IV. A New Synthesis of the Capsorubin Chromophore. J. Chem. Soc. (London) **1953**, 3815.
4. AKHTAR, M., T. A. RICHARDS and B. C. L. WEEDON: Studies with Acetylenes. III. The Synthesis of Three Partly-*cis*-Diphenyloctatetraenes. J. Chem. Soc. (London) **1959**, 933.
5. AKHTAR, M. and B. C. L. WEEDON: Carotenoids and Related Compounds. VIII. Novel Syntheses of Echinenone and Canthaxanthin. J. Chem. Soc. (London) **1959**, 4058.
6. ALDER, K. und M. SCHUMACHER: Anwendungen der Dien-Synthese für die Erforschung von Naturstoffen. Fortschr. Chem. organ. Naturstoffe **10**, 1 (1953).
7. ALLAN, J. L. H., G. D. MEAKINS and M. C. WHITING: Researches on Acetylenic Compounds. L. The Infrared Absorption of Some Conjugated Ethylenic and Acetylenic Systems. J. Chem. Soc. (London) **1955**, 1874.
8. AMES, D. E. and R. E. BOWMAN: Synthetic Long-chain Aliphatic Compounds. VIII. The Preparation of *cis*- and *trans*-Undec-9-enoic Acids. J. Chem. Soc. (London) **1952**, 677.
9. AMES, S. R.: Fat-soluble Vitamins. Annu. Rev. Biochem. **27**, 371 (1958).
10. AMES, S. R. and R. W. LEHMAN: Estimation of the Biological Potency of Vitamin A Sources from their Maleic Values. J. Assoc. Off. Agric. Chemists **43**, 21 (1960).
11. AMES, S. R., W. J. SWANSON and P. L. HARRIS: Biochemical Studies on Vitamin A. XIV. Biopotencies of Geometric Isomers of Vitamin A Acetate in the Rat. J. Amer. Chem. Soc. **77**, 4134 (1955).
12. — — — Biochemical Studies on Vitamin A. XV. Biopotencies of Geometric Isomers of Vitamin A Aldehyde in the Rat. J. Amer. Chem. Soc. **77**, 4136 (1955).
13. — — — In vivo Isomerization of Vitamin A Isomers. Federat. Proc. (Amer. Soc. exp. Biol.) **16**, 145 (1957).
14. AMES, S. R., W. J. SWANSON and R. W. LEHMAN: Estimation of the Biological Potency of Isomerized Vitamin A Palmitate in Aqueous Multivitamin Dispersions from Maleic Values. J. Amer. Pharm. Assoc., Sci. Ed. **49**, 366 (1960).
15. ANSCHÜTZ, R.: Über Fumar- und Maleinsäureäther. II. Ber. dtsch. chem. Ges. **12**, 2280 (1879).
16. ARENS, J. F. and D. A. VAN DORP: Synthesis of Vitamin A. Rec. trav. chim. Pays-Bas **68**, 604 (1949).
17. ARGOUD, S.: Oil-Palm Fruit Carotenoids. Oléagineux **13**, 249 (1958) [Chem. Abstr. **52**, 12425 (1958)].

18. BAEYER, A.: Über die Constitution des Benzols. Liebigs Ann. Chem. **245**, 103 (1888).

19. BAHL, A. N., J. C. SADANA and B. AHMAD: The Provitamin A Activities of Neo-β-carotene B and Neo-β-carotene U. J. Sci. Industr. Res. (India) **7 B**, 169 (1948).

20. BALL, S., T. W. GOODWIN and R. A. MORTON: Studies on Vitamin A. 5. The Preparation of Retinene$_1$ — Vitamin A Aldehyde. Biochemic. J. **42**, 516 (1948).

21. BARBER, M. S., J. B. DAVIS, L. M. JACKMAN and B. C. L. WEEDON: Studies in Nuclear Magnetic Resonance. Part I. Methyl Groups of Carotenoids and Related Compounds. J. Chem. Soc. (London) **1960**, 2870.

22. BARBER, M. S., L. M. JACKMAN, C. K. WARREN and B. C. L. WEEDON: The Structures of the Paprika Ketones. J. Chem. Soc. (London) **1960**, 19.

23. BARBER, M. S., L. M. JACKMAN and B. C. L. WEEDON: The Structures of Spirilloxanthin and Related Carotenoids. Proc. Chem. Soc. (London) **1959**, 96.

24. — — — Stereochemistry of Bixin. J. Chem. Soc. (London) **1960**, 23.

25. BARNHOLDT, B.: Separation of Neovitamin A$_1$ from All-*trans* Vitamin A$_1$ by Chromatography on Alumina. Nature (London) **178**, 1401 (1956).

26. — The Formation of Anhydro Vitamin A$_1$ in the Study of Vitamin A$_1$ Isomers. Acta Chem. Scand. **11**, 909 (1957).

27. BARNHOLDT, B. and W. HJARDE: Chromatographic Separation of two Vitamin A$_2$ Isomers. Acta Chem. Scand. **10**, 1635 (1956).

28. — — Chromatographic Separation of five Vitamin A$_1$ Isomers from the Eyes of Deep-Water Prawns, *Pandalus borealis*. Acta Physiol. Scand. **41**, 49 (1957) [Chem. Abstr. **52**, 4041 (1958)].

29. BAUER, L.: Trennung der Karotinoide und Chlorophylle mit Hilfe der Papierchromatographie. Naturwiss. **39**, 88 (1952).

30. BAXTER, J. G.: Synthesis and Properties of Vitamin A and Some Related Compounds. Fortschr. Chem. organ. Naturstoffe **9**, 41 (1952).

31. BEADLE, B. W. and F. P. ZSCHEILE: Studies on the Carotenoids. II. The Isomerization of β-Carotene and its Relation to Carotene Analysis. J. Biol. Chem. **144**, 21 (1942).

32. BERTHOD, H. et A. PULLMAN: Aspects de la structure électronique du rétinène et de ses isomères d'inérêt biologique. C. R. hebd. Séances Acad. Sci. **251**, 808 (1960).

33. BICKOFF, E. M.: Chromatographic Separation of β-Carotene Stereoisomers as a Function of Developing Solvent. Analyt. Chemistry **20**, 51 (1948).

34. BICKOFF, E. M., L. M. WHITE, A. BEVENUE and K. T. WILLIAMS: Isolation and Spectrophotometric Characterization of four Carotene Isomers. J. Assoc. Off. Agric. Chemists **31**, 633 (1948).

35. BLANKENHORN, D. H.: Carotenoids in Man. II. Fractions Obtained from Atherosclerotic and Normal Aortas, Serum, and Depot Fat by Separation on Alumina. J. Biol. Chem. **227**, 963 (1957).

36. BLISS, A. F.: Reversible Enzymic Reduction of Retinene to Vitamin A. Biol. Bull. **97**, 221 (1949).

37. BODEA, C. und E. NICOARA: Die Konstitution des Physoxanthins. Liebigs Ann. Chem. **609**, 181 (1957).

38. — — Zur Isolierung von α-Carotin, Physoxanthin und Lutein aus *Physalis Alkekengi*. Liebigs Ann. Chem. **622**, 188 (1959).

39. BOHLMANN, F. und H. J. MANNHARDT: Acetylenverbindungen im Pflanzenreich. Fortschr. Chem. organ. Naturstoffe **14**, 1 (1957).

40. BRAEKKAN, O. R., H. MYKLESTAD, L. R. NJAA and F. UTNE: Vitamin A Isomers in the Liver of Rats and Chicks. Nature (London) **186**, 312 (1960).

41. BRAUDE, E. A. and E. S. WAIGHT: The Relationships between the Stereochemistry and Spectroscopic Properties of Organic Compounds. Progr. Stereochem. I, 126 (1954). — Cf. E. A. BRAUDE: The Labile Stereochemistry of Conjugated Systems. Experientia II, 457 (1955).

42. BRINER, E. et E. DALLWIGK: Chaleurs d'ozonisation des isomères *cis* et *trans* des composés organiques à double liaison; applications à quelques oléfines, notamment au maléate et au fumarate d'éthyle. Helv. Chim. Acta 40, 2466 (1957).

43. BRODE, W. R.: Steric Effects in Dyes. In: The Roger Adams Symposium, p. 8. New York: Wiley, and London: Chapman & Hall. 1955.

44. BRO-RASMUSSEN, F., W. HJARDE and O. POROTNIKOFF: Chromatographic Separation of Vitamin-A-active Compounds in Cod-liver Oil. Analyst 80, 418 (1955).

45. BROWN, P. K. and G. WALD: The neo-b Isomer of Vitamin A and Retinene. J. Biol. Chem. 222, 865 (1955).

46. BROWN, P. S., W. P. BLUM and M. H. STERN: Isomers of Vitamin A in Fish Liver Oils. Nature (London) 184, 1377 (1959).

46 a. BRUBACHER, G., U. GLOOR und O. WISS: Zum Stoffwechsel von β-Apo-8'-carotinal (C_{30}). Chimia 14, 19 (1960).

47. BURWELL, R. L., Jr.: Stereochemistry and Heterogeneous Catalysis. Chem. Rev. 57, 895 (1957).

48. BUSH, W. V.: Thesis, California Institute of Technology, Pasadena, 1958.

49. BUSH, W. V. and L. ZECHMEISTER: On Some Cleavage Products of the Boron Trifluoride Complexes of α-Carotene, Lycopene, and γ-Carotene. J. Amer. Chem. Soc. 80, 2991 (1958).

50. CALVIN, M. and H. W. ALTER: Substituted Stilbenes. I. Absorption Spectra. J. Chem. Physics 19, 765 (1951).

51. — — Substituted Stilbenes. II. Thermal Isomerization. J. Chem. Physics 19, 768 (1951).

52. CAMA, H. R., F. D. COLLINS and R. A. MORTON: Studies in Vitamin A. 17. Spectroscopic Properties of all-*trans*-Vitamin A and Vitamin A Acetate. Analysis of Liver Oils. Biochemic. J. 50, 48 (1951).

53. CAMA, H. R., P. D. DALVI, R. A. MORTON and M. K. SALAH: Studies in Vitamin A. 21. Retinene₂ and Vitamin A₂. Biochemic. J. 52, 542 (1952).

54. CAMA, H. R., P. D. DALVI, R. A. MORTON, M. K. SALAH, G. R. STEINBERG and A. L. STUBBS: Studies in Vitamin A. 19. Preparation and Properties of Retinene₂. Biochemic. J. 52, 535 (1952).

55. CARTER, G. P. and A. E. GILLAM: The Isomerization of Carotenes. III. Reconsideration of the Change β-Carotene to $\psi\alpha$-Carotene. Biochemic. J. 33, 1325 (1939).

56. CAWLEY, J. D., C. D. ROBESON, L. WEISLER, E. M. SHANTZ, N. D. EMBREE and J. G. BAXTER: Crystalline Synthetic Vitamin A and Neovitamin A. Science (Washington) 107, 346 (1948).

57. CELMER, W. D. and I. A. SOLOMONS: Mycomycin. IV. Stereoisomeric 3,5-Diene Fatty Acid Esters. J. Amer. Chem. Soc. 75, 3430 (1953).

58. CHATAIN, H. et M. DEBODARD: A propos du dosage de la vitamine A. Spectre d'absorption du néo-axérophtol. C. R. hebd. Séances Acad. Sci. 232, 355 (1951).

59. CHATTERJEE, A. and L. ZECHMEISTER: On Some Stereoisomeric Cryptoxanthins. J. Amer. Chem. Soc. 72, 254 (1950).

60. CHOLNOKY, L., D. SZABÓ und J. SZABOLCS: Untersuchungen über die Carotinoidfarbstoffe. II. Die Struktur des Capsanthins und des Capsorubins. Liebigs Ann. Chem. 606, 194 (1957).

61. CHOLNOKY, L. und J. SZABOLCS: Untersuchungen über die Carotinoidfarb-stoffe. V. Über die Farbstoffe der Kelchblätter von *Physalis Alkekengi.* Liebigs Ann. Chem. **626**, 206 (1959).

62. CHOLNOKY, L., J. SZABOLCS und E. NAGY: Untersuchungen über die Carotinoid-farbstoffe. IV. α-Kryptoxanthin. Liebigs Ann. Chem. **616**, 207 (1958).

63. CLAES, H.: Biosynthese von Carotinoiden bei *Chlorella.* 2. Mitt. Tetrahydro-lycopin und Lycopin. Z. Naturforsch. **11 b**, 260 (1956).

64. — Biosynthese von Carotinoiden bei *Chlorella.* III. Untersuchungen über die lichtabhängige Synthese von α- und β-Carotin und Xanthophyllen bei der *Chlorella*-Mutante 5/520. Z. Naturforsch. **12 b**, 401 (1957).

64 a. — Interaction between Chlorophyll and Carotenes with Different Chromo-phoric Groups. Biochem. Biophys. Res. Comm. **3**, 585 (1960).

65. CLAES, H. and T. O. M. NAKAYAMA: Isomerization of Poly-*cis*-carotenes by Chlorophyll *in vivo* and *in vitro.* Nature (London) **183**, 1053 (1959).

66. CLARKE, P. M. and P. E. E. TODD: Some Observations on the Biological Assay of Vitamin A and its Precursors by the Vaginal-Smear Method. Brit. J. Nutrition **11**, 173 (1957).

67. COLE, A. R. H.: Infrared Spectra of Natural Products. Fortschr. Chem. organ. Naturstoffe **13**, 1 (1956).

68. COLLINS, F. D., J. N. GREEN and R. A. MORTON: Studies in Rhodopsin. 6. Regeneration of Rhodopsin. Biochemic. J. **53**, 152 (1953).

69. — — — Studies in Rhodopsin. 7. Regeneration of Rhodopsin by Comminuted Ox Retina. Biochemic. J. **56**, 493 (1954).

70. COOK, A. H.: Algal Pigments and their Significance. Biol. Rev. **20**, 115 (1945).

71. COOK, A. H. and D. G. JONES: *cis*-Azo-compounds. II. J. Chem. Soc. (London) **1939**, 1309.

72. COULSON, C. A.: The Electronic Structure of Some Polyenes and Aromatic Molecules. VII. Bonds of Fractional Order by the Molecular Orbital Method. Proc. Roy. Soc. (London), Ser. A **169**, 413 (1939).

73. CROMBIE, L.: Geometrical Isomerism about Carbon-Carbon Double Bonds. Quart. Rev. (Chem. Soc. London) **6**, 101 (1952).

74. CROMBIE, L., S. H. HARPER and R. J. D. SMITH: Stereochemical Studies of Olefinic Compounds. VI. Preparation of the Four Stereoisomeric Sorbyl Alcohols from a Single Precursor. J. Chem. Soc. (London) **1957**, 2754.

75. CURL, A. L.: Application of Countercurrent Distribution to Valencia Orange Juice Carotenoids. Agric. Food Chem. **1**, 456 (1953).

76. CURL, A. L. and G. F. BAILEY: Orange Carotenoids. Polyoxygen Carotenoids of Valencia Orange Juice. Agric. Food Chem. **2**, 685 (1954).

77. — — Orange Carotenoids. I. Comparison of Carotenoids of Valencia Orange Peel and Pulp. Agric. Food Chem. **4**, 156 (1956).

78. CURTIN, D. Y., H. GRUEN and B. A. SHOULDERS: Use of Nuclear Magnetic Resonance to Determine Configurations of *cis-trans* Isomers. Chem. and Ind. **1958**, 1205.

79. DALE, J.: Empirical Relationships of the Minor Bands in the Absorption Spectra of Polyenes. Acta Chem. Scand. **8**, 1235 (1954).

80. — The Free-Electron Model, "Overtone" Bands, and Vibrational Structure in Absorption Spectra of Polyenes and Polyenynes. Acta Chem. Scand. **11**, 265 (1957).

81. — Infrared Absorption Spectra of *ortho*- and *para*-Linked Polyphenyls. Acta Chem. Scand. **11**, 640 (1957).

82. — Ultraviolet Absorption Spectra of *ortho*- and *para*-Linked Polyphenyls. Acta Chem. Scand. **11**, 650 (1957).

83. DALE, J.: Ultraviolet Absorption Spectra of Chain Molecules Consisting of Alternating Benzene Rings and Ethylenic Bonds. Acta Chem. Scand. 11, 971 (1957).
84. DALE, J. and L. ZECHMEISTER: On the Stereochemistry of Azines: Cinnamalazine and Phenylpentadienalazine. J. Amer. Chem. Soc. 75, 2379 (1953).
85. DALVI, P. D. and R. A. MORTON: Studies in Vitamin A. 16. Preparation of Neovitamin A Esters and Neoretinene₁. Biochemic. J. 50, 43 (1951).
86. DARTNALL, H. J. A.: The Visual Pigments. New York: Wiley and Sons, and London: Methuen. 1957.
87. DAVIS, J. B., L. M. JACKMAN, P. T. SIDDONS and B. C. L. WEEDON: The Structures of Phytoene, Phytofluene, ζ-Carotene and Neurosporene. J. Chem. Soc. (London) 1961, 261.
88. DEUEL, H. J., Jr.: The Lipids. Vol. III: Biochemistry, p. 540. New York: Interscience Publ. 1957.
89. DEUEL, H. J., Jr., J. GANGULY, B. K. KOE and L. ZECHMEISTER: Stereoisomerization of the Polycis Compounds, Pro-γ-carotene and Prolycopene in Chickens and Hens. Arch. Biochem. Biophys. 33, 143 (1951).
90. DEUEL, H. J., Jr., S. M. GREENBERG, E. STRAUB, T. FUKUI, A. CHATTERJEE and L. ZECHMEISTER: Stereochemical Configuration and Provitamin A Activity. VII. Neocryptoxanthin U. Arch. Biochemistry 23, 239 (1949).
91. DEUEL, H. J., Jr., C. HENDRICK, E. STRAUB, A. SANDOVAL, J. H. PINCKARD and L. ZECHMEISTER: Stereochemical Configuration and Provitamin A Activity. VI. Some cis-trans Isomers of γ-Carotene. Arch. Biochemistry 14, 97 (1947).
92. DEUEL, H. J., Jr., H. H. INHOFFEN, J. GANGULY, L. WALLCAVE and L. ZECHMEISTER: Stereochemical Configuration and Provitamin A Activity. XI. A Comparison of Synthetic All-trans- and Dicis-16,16'-homo-β-carotene C₄₂H₅₈ with All-trans-β-carotene in the Rat. Arch. Biochem. Biophys. 40, 352 (1952).
93. DEUEL, H. J., Jr., C. H. JOHNSTON, E. R. MESERVE, A. POLGÁR and L. ZECHMEISTER: Stereochemical Configuration and Provitamin A Activity. IV. Neo-α-carotene B and Neo-β-carotene B. Arch. Biochemistry 7, 247 (1945).
94. DEUEL, H. J., Jr., C. H. JOHNSTON, E. SUMNER, A. POLGÁR, W. A. SCHROEDER and L. ZECHMEISTER: Stereochemical Configuration and Provitamin A Activity. II. All-trans-γ-carotene and Pro-γ-carotene. Arch. Biochemistry 5, 365 (1944).
95. DEUEL, H. J., Jr., C. H. JOHNSTON, E. SUMNER, A. POLGÁR and L. ZECHMEISTER: Stereochemical Configuration and Provitamin A Activity. I. All-trans-β-carotene and Neo-β-carotene U. Arch. Biochemistry 5, 107 (1944).
96. DEUEL, H. J., Jr., E. R. MESERVE, A. SANDOVAL and L. ZECHMEISTER: Stereochemical Configuration and Provitamin A Activity. V. Neocryptoxanthin A. Arch. Biochemistry 10, 491 (1946).
97. DEUEL, H. J., Jr., E. SUMNER, C. H. JOHNSTON, A. POLGÁR and L. ZECHMEISTER: Stereochemical Configuration and Provitamin A Activity. III. All-trans-α-carotene and Neo-α-carotene U. Arch. Biochemistry 6, 157 (1945).
98. DIETERLE, J. M. and C. D. ROBESON: Crystalline Neoretinene b. Science (Washington) 120, 219 (1954).
99. DREFAHL, G. und G. PLÖTNER: Untersuchungen über Stilbene. XX. Polyphenyl-polybutadiene. Chem. Ber. 91, 1285 (1958).
100. ECKERT, R. und H. KUHN: Richtungen der Übergangsmomente der Absorptionsbanden von Polyenen, Cyaninen und Vitamin B₁₂ aus Dichroismus und Fluoreszenzpolarisation. Z. Elektrochem. 64, 356 (1960).
101. EDISBURY, J. R., A. E. GILLAM, I. M. HEILBRON and R. A. MORTON: Absorption Spectra of Substances Derived from Vitamin A. Biochemic. J. 26, 1164 (1932).
102. ELVIDGE, J. A., R. P. LINSTEAD and P. SIMS: Polyene Acids. III. A Reinvestigation of Karrer's β-Methylmuconic Acid. J. Chem. Soc. (London) 1951, 3398.

103. EMBREE, N. D., S. R. AMES, R. W. LEHMAN and P. L. HARRIS: Determination of Vitamin A. Methods Biochem. Anal. **4**, 43 (1957).

104. ENTSCHEL, R. und P. KARRER: Zur Konstitution des Capsanthins und Capsorubins. Helv. Chim. Acta **43**, 89 (1960).

105. ERDTMAN, H.: Corticrocin, a Pigment from the Mycelium of a *Mycorrhiza* Fungus. II. Acta Chem. Scand. **2**, 209 (1948).

106. EUGSTER, C. H., C. (F.) GARBERS und P. KARRER: Carotinoidsynthesen. IX. Stereoisomere 1,18-Diphenyl-3,7,12,16-tetramethyl-octadeca-nonaene. Helv. Chim. Acta **35**, 1179 (1952).

107. — — — Carotinoidsynthesen. XIII. Über zwei isomere *cis-β*-Carotine mit *cis*-Konfiguration an „behinderten" Doppelbindungen. Helv. Chim. Acta **36**, 1378 (1953).

108. EUGSTER, C. H. und P. KARRER: Taraxanthin und Tarachrom sowie Beobachtungen über stereoisomere Trollixanthine. Helv. Chim. Acta **40**, 69 (1957).

109. EUGSTER, C. H., E. LINNER, A. H. TRIVEDI und P. KARRER: Carotinoidsynthesen. XIX. Synthese eines 6,7,6′,7′-Tetrahydro-lycopins und dessen Beziehung zum Neurosporin. Helv. Chim. Acta **39**, 690 (1956).

110. EUGSTER, C. H., A. H. TRIVEDI und P. KARRER: Carotinoidsynthesen. XVII. Synthese des 2,2′-Dimethyl-β-carotins. Helv. Chim. Acta **38**, 1359 (1955).

111. EULER, B. v., H. v. EULER und H. HELLSTRÖM: A-Vitaminwirkungen der Lipochrome. Biochem. Z. **203**, 370 (1928).

112. EULER, B. v., H. v. EULER und P. KARRER: Zur Biochemie der Carotinoide. Helv. Chim. Acta **12**, 278 (1929).

113. EULER, H. v., P. KARRER und U. SOLMSSEN: Homologe des Vitamins A (Axerophtols) und ein Abbauprodukt des α-Carotins, α-Apo-2-carotinal. Helv. Chim. Acta **21**, 211 (1938).

113 a. FAIGLE, H. und P. KARRER: Konstitution und Konfiguration von Capsanthin und Capsorubin. Helv. Chim. Acta **44**, 1257 (1961).

114. FARKAS, A. and L. FARKAS: The Mechanism of Hydrogenation Reactions and the Formation of Stereochemical Isomers. Trans. Faraday Soc. **33**, 837 (1937).

115. FARRAR, K. R., J. C. HAMLET, H. B. HENBEST and E. R. H. JONES: Studies in the Polyene Series. XLIII. The Structure and Synthesis of Vitamin A_2 and Related Compounds. J. Chem. Soc. (London) **1952**, 2657.

116. FAZAKERLEY, S. and J. GLOVER: The Provitamin A Activity of Some Possible Intermediates in β-Carotene Metabolism. Biochemic. J. **65**, 38 P (1957).

117. FISHER, L. R. and S. K. KON: Vitamin A in the Invertebrates. Biol. Rev. Cambridge Philos. Soc. **34**, 1 (1959).

118. FISHER, L. R., S. K. KON and P. A. PLACK: Vitamin A Isomers in some Eucaridan Crustacea. Proc. Roy. Soc. B **147**, 200 (1957).

119. FISHWICK, M. J. and J. GLOVER: The Metabolism of Uniformly ^{14}C-labelled β-Carotene in the Rat. Biochemic. J. **66**, 36 P (1957).

120. FORBES, W. F. and R. SHILTON: Electronic Spectra and Molecular Dimensions. III. Steric Effects in Methyl-substituted α,β-Unsaturated Aldehydes. J. Amer. Chem. Soc. **81**, 786 (1959).

121. FRANK, R. L., R. D. EMMICK and R. S. JOHNSON: *cis-* and *trans*-Piperylenes. J. Amer. Chem. Soc. **69**, 2313 (1947).

122. GANSSER, CH. and L. ZECHMEISTER: On Some *cis*-Forms of Phenylundecapentaenal. J. Amer. Chem. Soc. **79**, 3854 (1957).

123. — — Über einige *cis*-Formen des Canthaxanthins. Helv. Chim. Acta **40**, 1757 (1957).

124. GARBERS, C. F.: Synthesis of all-*trans*-, 2-*cis*- (or *neo*-), and 6-*cis*-[2-^{14}C] Vitamin A. J. Chem. Soc. (London) **1956**, 3234.

125. GARBERS, C. F., C. H. EUGSTER und P. KARRER: Carotinoidsynthesen.
X. Weitere stereoisomere 1,18-Diphenyl-3,7,12,16-tetramethyl-octadecanonaene.
Zugleich ein Beitrag zu L. Pauling's Theorie der sterischen Hinderung bei
cis-trans-isomeren Polyenen. Helv. Chim. Acta 35, 1850 (1952).

126. — — — Carotinoidsynthesen. XI. Weitere, mit 1,18-Diphenyl-3,7,12,16-
tetramethyl-octadecanonaen verwandte Polyene. Helv. Chim. Acta 36, 562
(1953).

127. — — — Über die Vitamin-A-Wirkung des cis-β-Carotins C (mit behinderter
Doppelbindung). Helv. Chim. Acta 37, 382 (1954).

128. GARBERS, C. F. und P. KARRER: Carotinoidsynthesen. XII. Bis-dehydro-
lycopine und totalsynthetische cis-Lycopine. Helv. Chim. Acta 36, 828
(1953).

129. GILLAM, A. E. and M. S. EL RIDI: Adsorption of Grass and Butter Carotenes
on Alumina. Nature (London) 136, 914 (1935).

130. — — The Isomerization of Carotenes by Chromatographic Adsorption.
I. Pseudo-α-carotene. Biochemic. J. 30, 1735 (1936).

131. GILLAM, A. E., M. S. EL RIDI and S. K. KON: The Isomerization of Carotenes
by Chromatographic Adsorption. II. Neo-α-carotene. Biochemic. J. 31, 1605
(1937).

131 a. GLOVER, J.: The Conversion of β-Carotene into Vitamin A. Vitamins and
Horm. 18, 371 (1960).

132. GLOVER, J., T. W. GOODWIN and R. A. MORTON: Studies in Vitamin A.
6. Conversion in vivo of Vitamin A Aldehyde (Retinene₁) to Vitamin A₁.
Biochemic. J. 43, 109 (1948).

133. — — — Studies in Vitamin A. 8. Conversion of β-Carotene into Vitamin A
in the Intestine of the Rat. Biochemic. J. 43, 512 (1948).

134. GLOVER, J. and E. R. REDFEARN: The Mechanism of the Transformation
of β-Carotene into Vitamin A in vivo. Biochemic. J. 58, xv (1954).

135. GOODWIN, T. W.: The Comparative Biochemistry of the Carotenoids. London:
Chapman & Hall. 1952.

136. — Studies in Carotenogenesis. 13. The Carotenoids of the Flower Petals of
Calendula officinalis. Biochemic. J. 58, 90 (1954).

137. — Carotenoids. Annu. Rev. Biochem. 24, 497 (1955).

138. — Studies in Carotenogenesis. 19. A Survey of the Polyenes in a Number
of Ripe Berries. Biochemic. J. 62, 346 (1956).

139. GOODWIN, T. W. and M. JAMIKORN: Studies in Carotenogenesis. 11. Carotenoid
Synthesis in the Alga Haematococcus pluvialis. Biochemic. J. 57, 376 (1954).

140. GOODWIN, T. W. and D. G. LAND: Studies in Carotenogenesis. 20. Carotenoids
of Some Species of Chlorobium. Biochemic. J. 62, 553 (1956).

141. GRAHAM, W., D. A. VAN DORP and J. F. ARENS: Synthesis of a cis Isomer
of Vitamin A Aldehyde. Rec. trav. chim. Pays-Bas 68, 609 (1949).

142. GRANGAUD, R. et P. CHARDENOT: Stéréoisomérisation cis-trans de l'astaxanthine. C. R. hebd. Séances Acad. Sci. 242, 1767 (1956).

143. GRANGAUD, R. et I. GARCIA: Chromatographie de partage des caroténoïdes.
I. Séparation de l'astaxanthine. Bull. soc. chim. biol. (Paris) 34, 754
(1952).

144. GREEN, A., C. H. EUGSTER und P. KARRER: Über Inhaltsstoffe der Wurzeln
von Aristolochia cymbifera MART. Helv. Chim. Acta 37, 1717 (1954).

145. GREENBERG, S. M., C. E. CALBERT, J. H. PINCKARD, H. J. DEUEL, Jr. and
L. ZECHMEISTER: Stereochemical Configuration and Provitamin A Activity.
IX. A Comparison of All-trans-γ-carotene and Pro-γ-carotene with All-trans-
β-carotene in the Chick. Arch. Biochemistry 24, 31 (1949).

146. GREENBERG, S. M., A. CHATTERJEE, C. E. CALBERT, H. J. DEUEL, Jr. and L. ZECHMEISTER: A Comparison of the Provitamin A Activity of β-Carotene and Cryptoxanthin in the Chick. Arch. Biochemistry **25**, 61 (1950).

147. GRUMMITT, O. and F. J. CHRISTOPH: Geometric Isomers of 1-Phenyl-1,3-butadiene. J. Amer. Chem. Soc. **71**, 4157 (1949).

148. — — Geometric Isomers of 1-Phenyl-1,3-butadiene. J. Amer. Chem. Soc. **73**, 3479 (1951).

149. HAAGEN-SMIT, A. J., J. H. PINCKARD and L. ZECHMEISTER: Contribution to the Structure of Pro-γ-carotene and Prolycopene Obtained from Various Sources. Arch. Biochemistry **26**, 358 (1950).

150. HARMAN, R. A. and H. EYRING: The Structure of Substituted Ethylenes and their Isomerization, Polymerization and "Peroxide Addition" Reactions. J. Chem. Physics **10**, 557 (1942).

151. HARRIS, P. L., S. R. AMES and J. H. BRINKMAN: Biochemical Studies on Vitamin A. IX. Biopotency of Neovitamin A in the Rat. J. Amer. Chem. Soc. **73**, 1252 (1951).

152. HASSELT, J. F. B. VAN: Études sur la constitution de la bixine. Rec. trav. chim. Pays-Bas **30**, 1 (1911).

153. HAUSSER, K. W., R. KUHN und G. SEITZ: Lichtabsorption und Doppelbindung. V. Über die Absorption von Verbindungen mit konjugierten Kohlenstoffdoppelbindungen bei tiefer Temperatur. Z. physik. Chem. B **29**, 391 (1935).

154. HAXO, F. (T.): Studies on the Carotenoid Pigments of *Neurospora*. I. Composition of the Pigment. Arch. Biochemistry **20**, 400 (1949).

155. HAYES, E. and M. A. PETITPIERRE: Neovitamin A and Vitamin A Alcohol in Commercial Fish Liver Oils and Vitamin A Concentrates. J. Pharm. Pharmacol. **4**, 879 (1952).

156. HEILBRON, I. M., R. N. HESLOP, R. A. MORTON, E. T. WEBSTER, J. L. REA and J. C. DRUMMOND: Characteristics of Highly Active Vitamin A Preparations. Biochemic. J. **26**, 1178 (1932).

157. HEILBRON, I. M., A. W. JOHNSON, E. R. H. JONES and A. SPINKS: Studies in the Polyene Series. Part V. The Employment of 3-(2':6':6'-Trimethyl-*cyclo*hexenyl)-1-methylcrotonaldehyde for the Synthesis of Vitamin A and Analogues. J. Chem. Soc. (London) **1942**, 727.

158. HEILBRON, I. M., R. A. MORTON and E. T. WEBSTER: The Structure of Vitamin A. Biochemic. J. **26**, 1194 (1932).

159. HENBEST, H. B., E. R. H. JONES and T. C. OWEN: Studies in the Polyene Series. Part LI. Conversion of Vitamin A_1 into Vitamin A_2. J. Chem. Soc. (London) **1955**, 2765.

160. HENGSTENBERG, J. und R. KUHN: Die Kristallstruktur der Diphenylpolyene. Z. Kristallogr., Mineral., Petrogr. A **75**, 301 (1930).

161. HERZIG, J. und F. FALTIS: Zur Kenntnis des Bixins. Liebigs Ann. Chem. **431**, 40 (1923).

162. HIRSHBERG, Y., E. BERGMANN and F. BERGMANN: The Fluorescence of 1,4-Diarylbutadienes and its Relation to their Configuration. J. Amer. Chem. Soc. **72**, 5117 (1950).

163. — — — Absorption Spectra and Photo-isomerization of Arylated Dienes. J. Amer. Chem. Soc. **72**, 5120 (1950).

164. HOLME, D., E. R. H. JONES and M. C. WHITING: The Synthesis of an All-*cis*-tetraene. Chem. and Ind. **1956**, 928.

165. HUBBARD, R.: *Cis-trans* Isomers of Vitamin A and Retinene in the Rhodopsin System. Federat. Proc. (Amer. Soc. exp. Biol.) **11**, 233 (1952).

166. — Retinene Isomerase. Federat. Proc. (Amer. Soc. exp. Biol.) **14**, 229 (1955).

167. HUBBARD, R.: Geometrical Isomerization of Vitamin A, Retinene and Retinene Oxime. J. Amer. Chem. Soc. **78**, 4662 (1956).
168. — Retinene Isomerase. J. Gen. Physiol. **39**, 935 (1956).
169. — Bleaching of Rhodopsin by Light and by Heat. Nature (London) **181**, 1126 (1958).
170. — The Thermal Stability of Rhodopsin and Opsin. J. Gen. Physiol. **42**, 259 (1958).
171. HUBBARD, R., P. K. BROWN and A. KROPF: Action of Light on Visual Pigments. Vertebrate Lumi- and Meta-rhodopsins. Nature (London) **183**, 442 (1959).
172. HUBBARD, R. and A. D. COLMAN: Vitamin-A Content of the Frog Eye during Light and Dark Adaptation. Science (Washington) **130**, 977 (1959).
173. HUBBARD, R., R. I. GREGERMAN and G. WALD: Geometrical Isomers of Retinene. J. Gen. Physiol. **36**, 415 (1953).
174. HUBBARD, R. and A. KROPF: The Action of Light on Rhodopsin. Proc. Nat. Acad. Sci. (USA) **44**, 130 (1958).
175. — — Chicken Lumi- and Meta-iodopsin. Nature (London) **183**, 448 (1959).
176. HUBBARD, R. and G. WALD: The Mechanism of Rhodopsin Synthesis. Proc. Nat. Acad. Sci. (USA) **37**, 69 (1951).
177. — — *Cis-trans* Isomers of Vitamin A and Retinene in Vision. Science (Washington) **115**, 60 (1952).
178. — — *Cis-trans* Isomers of Vitamin A and Retinene in the Rhodopsin System. J. Gen. Physiol. **36**, 269 (1952).
179. HUISMAN, H. O., A. SMIT, P. H. VAN LEEUWEN and J. H. VAN RIJ: Investigations in the Vitamin A Series. III. Rearrangement of the *retro*-System to the *normal* System of Conjugated Double Bonds in the Vitamin A Series. Rec. trav. chim. Pays-Bas **75**, 977 (1956).
180. HUNTER, R. F. and E. G. E. HAWKINS: Vitamin A Aldehyde. Nature (London) **153**, 194 (1944).
181. HUNTER, R. F., A. D. SCOTT and J. R. EDISBURY: Palm Oil Carotenoids. 2. The Isolation of Lipoid Pigments from a West African Plantation Oil and Some Remarks on the Isomerization of Carotenoids. Biochemic. J. **36**, 697 (1942).
182. INHOFFEN, H. H. und G. VON DER BEY: Synthesen in der Carotinoid-Reihe. XXVII. Stereoisomere C_{10}-Diole und Dialdehyde. Liebigs Ann. Chem. **583**, 100 (1953).
183. INHOFFEN, H. H., F. BOHLMANN, H.-J. ALDAG, S. BORK und G. LEIBNER: Synthesen in der Carotinoid-Reihe. XXI. Kondensation von Carotinoid-ketonen und -aldehyden mit Diacetylen; zugleich eine weitere Synthese des β-Carotins. Liebigs Ann. Chem. **573**, 1 (1951).
184. INHOFFEN, H. H., F. BOHLMANN, K. BARTRAM und H. POMMER: Totalsynthese des β-Carotins. Chem.-Ztg. **74**, 285 (1950).
185. INHOFFEN, H. H., F. BOHLMANN, K. BARTRAM, G. RUMMERT und H. POMMER: Synthesen in der Carotinoid-Reihe. XV. Über die Darstellung von *trans* und von 9,9'-mono-*cis*-β-Carotin. Liebigs Ann. Chem. **570**, 54 (1950).
186. INHOFFEN, H. H., F. BOHLMANN und G. RUMMERT: Synthesen in der Carotinoid-Reihe. X. 7,7'-Bis-desmethyl-β-carotin. Liebigs Ann. Chem. **569**, 226 (1950).
187. — — — Synthesen in der Carotinoid-Reihe. XVIII. Über die Stereo-isomerisierung des 9,9'-mono-*cis*-β-Carotins. Liebigs Ann. Chem. **571**, 75 (1951).
188. INHOFFEN, H. H., O. ISLER, G. VON DER BEY, G. RASPÉ, P. ZELLER und R. AHRENS: Synthesen in der Carotinoid-Reihe. XXVI. Totalsynthese des Crocetin-dimethylesters. Liebigs Ann. Chem. **580**, 7 (1953).

189. INHOFFEN, H. H., H.-J. KRAUSE und S. BORK: Synthesen in der Carotinoid-Reihe. XXIX. Liebigs Ann. Chem. 585, 132 (1953).

190. INHOFFEN, H. H. und G. RASPÉ: Synthesen in der Carotinoid-Reihe. XXXI. Synthese des 10,10'-cis-Bixinmethylesters. Ein Beitrag zur Stereochemie der cis-trans-isomeren Bixine. Liebigs Ann. Chem. 592, 214 (1955).

191. — — Synthesen in der Carotinoid-Reihe. XXXII. Totalsynthese des 3,4,3',4'-Bisdehydro-β-carotins. Liebigs Ann. Chem. 594, 165 (1955).

192. INHOFFEN, H. H., U. SCHWIETER und G. RASPÉ: Synthesen in der Carotinoid-Reihe. XXX. Totalsynthese des d,l-α-Carotins. Liebigs Ann. Chem. 588, 117 (1954).

193. INHOFFEN, H. H. und H. SIEMER: Synthetische Chemie der Carotinoide. Fortschr. Chem. organ. Naturstoffe 9, 1 (1952).

194. ISLER, O.: Über das Vitamin A und die Carotinoide. Angew. Chem. 68, 547 (1956).

195. ISLER, O., L. H. CHOPARD-DIT-JEAN, M. MONTAVON, R. RÜEGG und P. ZELLER: Synthesen in der Carotinoid-Reihe. 12. Mitt. Synthese von 11,11'-Di-cis-β-carotin nach einem neuen Aufbauprinzip. Helv. Chim. Acta 40, 1256 (1957).

196. ISLER, O., W. GUEX, R. RÜEGG, G. RYSER, G. SAUCY, U. SCHWIETER, M. WALTER und A. WINTERSTEIN: Synthesen in der Carotinoid-Reihe. 16. Mitt. Carotinoide vom Typus des Torularhodins. Helv. Chim. Acta 42, 864 (1959).

197. ISLER, O., H. GUTMANN, H. LINDLAR, M. MONTAVON, R. RÜEGG, G. RYSER und P. ZELLER: Synthesen in der Carotinoid-Reihe. 6. Mitt. Synthese von Crocetindialdehyd und Lycopin. Helv. Chim. Acta 39, 463 (1956).

198. ISLER, O., H. GUTMANN, M. MONTAVON, R. RÜEGG, G. RYSER und P. ZELLER: Synthesen in der Carotinoid-Reihe. 10. Mitt. Anwendung der Wittig-Reaktion zur Synthese von Estern des Bixins und Crocetins. Helv. Chim. Acta 40, 1242 (1957).

199. ISLER, O., W. HUBER, A. RONCO und M. KOFLER: Synthese des Vitamin A. Helv. Chim. Acta 30, 1911 (1947).

200. ISLER, O., M. KOFLER, W. HUBER und A. RONCO: Synthese von Vitamin-A-Methyläther. Experientia 2, 31 (1946).

201. ISLER, O., H. LINDLAR, M. MONTAVON, R. RÜEGG, G. SAUCY und P. ZELLER: Synthesen in der Carotinoid-Reihe. 7. Mitt. Totalsynthese von Zeaxanthin und Physalien. Helv. Chim. Acta 39, 2041 (1956).

202. — — — — — — Synthesen in der Carotinoid-Reihe. 8. Mitt. Totalsynthese von Kryptoxanthin und eine weitere Synthese von Zeaxanthin. Helv. Chim. Acta 40, 456 (1957).

203. ISLER, O., H. LINDLAR, M. MONTAVON, R. RÜEGG und P. ZELLER: Synthesen in der Carotinoid-Reihe. 1. Mitt. Die technische Synthese von β-Carotin. Helv. Chim. Acta 39, 249 (1956).

204. — — — — — Synthesen in der Carotinoid-Reihe. 3. Mitt. Die Synthese von 3,4;3',4'-Bisdehydro-β-carotin und 3,4-Monodehydro-β-carotin. Helv. Chim. Acta 39, 274 (1956).

205. — — — — — Synthesen in der Carotinoid-Reihe. 4. Mitt. Synthese von Isozeaxanthin. Helv. Chim. Acta 39, 449 (1956).

206. ISLER, O., M. MONTAVON, R. RÜEGG, G. SAUCY und P. ZELLER: Synthese hydroxylhaltiger Carotinoide. Verh. naturforsch. Ges. Basel 67, 379 (1956).

207. ISLER, O., M. MONTAVON, R. RÜEGG und P. ZELLER: Synthesen in der Carotinoid-Reihe. 9. Mitt. Neuer Aufbau symmetrischer Carotinoide. Liebigs Ann. Chem. 603, 129 (1957).

208. ISLER, O., A. RONCO, W. GUEX, N. C. HINDLEY, W. HUBER, K. DIALER und M. KOFLER: Über die Ester und Äther des synthetischen Vitamins A. Helv. Chim. Acta **32**, 489 (1949).

209. ISLER, O., R. RÜEGG und U. SCHWIETER: Sterisch gehinderte Raumformen der Vitamine A und A$_2$. Chimia **15**, 288 (1961); cf. **14**, 362 (1960).

209 a. ISLER, O., R. RÜEGG, U. SCHWIETER and J. WÜRSCH: The Synthesis and Labeling of Vitamin A and Related Compounds. Vitamins and Horm. **18**, 295 (1960).

210. ISLER, O. and P. ZELLER: Total Syntheses of Carotenoids. Vitamins and Horm. **15**, 31 (1957).

211. JACKSON, J. E., R. F. PASCHKE, W. TOLBERG, H. M. BOYD and D. H. WHEELER: Isomers of Linoleic Acid. Infrared and Ultraviolet Properties of Methyl Esters. J. Amer. Oil Chem. Soc. **29**, 229 (1952).

212. JENKINS, J. A. and G. MACKINNEY: Inheritance of Carotenoid Differences in the Tomato Hybrid Yellow × Tangerine. Genetics **38**, 107 (1953).

213. JENSEN, A. and S. L. JENSEN: Quantitative Paper Chromatography of Carotenoids. Acta Chem. Scand. **13**, 1863 (1959).

214. JENSEN, S. L.: A Note on the Constitutions of Spirilloxanthin and P-481. Acta Chem. Scand. **13**, 381 (1959).

215. JENSEN, S. L., G. COHEN-BAZIRE, T. O. M. NAKAYAMA and R. Y. STANIER: The Path of Carotenoid Synthesis in a Photosynthetic Bacterium. Biochim. Biophys. Acta **29**, 477 (1958).

216. JOHNSON, R. M. and C. A. BAUMANN: Storage and Distribution of Vitamin A in Rats Fed Certain Isomers of Carotene. Arch. Biochemistry **14**, 361 (1947).

217. JONES, R. N.: The Ultraviolet Absorption Spectra of Arylethylenes. J. Amer. Chem. Soc. **65**, 1818 (1943).

218. JOYCE, A. E.: Some Polyenes from *Brassica rutabaga*. Nature (London) **173**, 311 (1954).

219. — Carotenoids of *Brassica napus*. J. Sci. Food Agric. **10**, 342 (1959).

220. JURKOWITZ, L.: Photochemical and Stereochemical Properties of Carotenoids at Low Temperatures. I. Photochemical Behaviour of Retinene. Nature (London) **184**, 614 (1959).

221. KARMAKAR, G. and L. ZECHMEISTER: On Some Dehydrogenation Products of α-Carotene, β-Carotene and Cryptoxanthin. J. Amer. Chem. Soc. **77**, 55 (1955).

222. KARRER, P.: Carotinoid-epoxyde und furanoide Oxyde von Carotinoidfarbstoffen. Fortschr. Chem. organ. Naturstoffe **5**, 1 (1948).

223. — Syntheses and Stereochemistry of Carotenoids. J. Sci. Industr. Res. (India), Ser. A **14**, 166 (1955).

224. — Zur Konstitution des Rhodoviolascins (Spirilloxanthin). Helv. Chim. Acta **43**, 181 (1960).

225. KARRER, P. und J. BENZ: Axerophten, der dem Vitamin A zugrunde liegende Kohlenwasserstoff. Helv. Chim. Acta **31**, 1048 (1948).

226. KARRER, P. et C. H. EUGSTER: Synthèse totale du β-carotène. C. R. hebd. Séances Acad. Sci. **230**, 1920 (1950).

227. — — Synthese von Carotinoiden. II. Totalsynthese des β-Carotins. I. Helv. Chim. Acta **33**, 1172 (1950).

228. — — Carotinoidsynthesen. VIII. Synthese des Dodecapreno-β-carotins. Helv. Chim. Acta **34**, 1805 (1951).

229. KARRER, P., C. H. EUGSTER und M. FAUST: Über das Auftreten von Carotinoiden in Pollen und Staubbeuteln verschiedener Pflanzen. Helv. Chim. Acta **33**, 300 (1950).

230. KARRER, P., A. HELFENSTEIN, R. WIDMER und TH. B. VAN ITALLIE: Über Bixin. XIII. Mitt. über Pflanzenfarbstoffe. Helv. Chim. Acta **12**, 741 (1929).

231. KARRER, P. und W. HESS: Über die katalytische Oxydation von Vitamin A mit Sauerstoff und Platin zu Vitamin-A-aldehyd (eine neue Methode). Helv. Chim. Acta **40**, 265 (1957).

232. KARRER, P. and E. JUCKER: Carotenoids. Translated and Revised by E. A. BRAUDE. New York: Elsevier Publ. Co. 1950.

233. KARRER, P., R. MORF und K. SCHÖPP: Zur Kenntnis des Vitamins A aus Fischtranen. II. Helv. Chim. Acta **14**, 1431 (1931).

234. — — — Synthese des Perhydro-vitamins A. Helv. Chim. Acta **16**, 557 (1933).

235. KARRER, P. und J. RUTSCHMANN: Über Violaxanthin, Auroxanthin und andere Pigmente der Blüten von *Viola tricolor.* Helv. Chim. Acta **27**, 1684 (1944).

236. — — Carotinoide aus den Früchten von *Cotoneaster occidentalis* und *Pyracantha coccinia.* Helv. Chim. Acta **28**, 1528 (1945).

237. KARRER, P., R. SCHWYZER und A. NEUWIRTH: Oxydation von 4-Methyl-o-benzochinon zu cis-cis-β-Methylmuconsäure-anhydrid. Helv. Chim. Acta **31**, 1210 (1948).

238. KARRER, P. und U. SOLMSSEN: Stufenweiser Abbau des labilen und stabilen Bixins. Zur Stereochemie der Carotinoide. Helv. Chim. Acta **20**, 1396 (1937).

239. KARRER, P. und T. TAKAHASHI: Pflanzenfarbstoffe. XLVII. Über die Isomerieverhältnisse beim Bixin. Bemerkungen zu den Theorien über die Bildung von Carotinoidpigmenten in der Pflanze. Helv. Chim. Acta **16**, 287 (1933).

240. KARRER, P. und E. WÜRGLER: Absorptionsspektren einiger Carotinoide. Helv. Chim. Acta **26**, 116 (1943).

241. KAWAKAMI, K.: On New Crystalline Derivatives of Vitamin A. Sci. Papers Inst. Phys. Chem. Res. (Tokyo) **26**, 77 (1935).

242. KELBER, C. und A. SCHWARZ: Über kolloidales Palladium. Partielle und totale Hydrogenisation von Phenyl-acetylen, Tolan und Diphenyl-diacetylen. Ber. dtsch. chem. Ges. **45**, 1946 (1912).

243. KEMMERER, A. R. and G. S. FRAPS: Constituents of Carotene Extracts of Plants. Ind. Eng. Chem., Analyt. Ed. **15**, 714 (1943).

244. — — Relative Value of Carotene in Vegetables for Growth of the White Rat. Arch. Biochemistry **8**, 197 (1945).

245. — — The Vitamin A Activity of neo-β-Carotene U and its Steric Rearrangement in the Digestive Tract of Rats. J. Biol. Chem. **161**, 305 (1945).

246. KHARASCH, M. S., J. V. MANSFIELD and F. R. MAYO: cis-trans Isomerization by Bromine Atoms. J. Amer. Chem. Soc. **59**, 1155 (1937).

247. KISTIAKOWSKY, G. B. and W. R. SMITH: Kinetics of Thermal cis-trans Isomerization. III. J. Amer. Chem. Soc. **56**, 638 (1934).

248. KODICEK, E.: Fat-soluble Vitamins. Annu. Rev. Biochem. **25**, 497 (1956).

249. KOE, B. K. and L. ZECHMEISTER: In Vitro Conversion of Phytofluene and Phytoene into Carotenoid Pigments. Arch. Biochem. Biophys. **41**, 236 (1952).

250. — — Preparation and Spectral Characteristics of all-*trans*- and a cis-Phytofluene. Arch. Biochem. Biophys. **46**, 100 (1953).

250 a. KOFLER, M. and S. H. RUBIN: Physicochemical Assay of Vitamin A and Related Compounds. Vitamins and Horm. **18**, 315 (1960).

251. KOHLER, E. P., J. T. WALKER and M. TISHLER: The Resolution of an Allenic Compound. J. Amer. Chem. Soc. **57**, 1743 (1935).

252. KRINSKY, N. I.: The Enzymatic Esterification of Vitamin A. J. Biol. Chem. **232**, 881 (1958).

253. KROPF, A. and R. HUBBARD: The Mechanism of Bleaching Rhodopsin. Ann. New York Acad. Sci. 74, 266 (1958).
254. KUHN, H.: The Electron Gas Theory of the Color of Natural and Artificial Dyes: Problems and Principles. Fortschr. Chem. organ. Naturstoffe 16, 169 (1958).
255. — The Electron Gas Theory of the Color of Natural and Artificial Dyes: Applications and Extensions. Fortschr. Chem. organ. Naturstoffe 17, 404 (1959).
256. — Neuere Untersuchungen über das Elektronengasmodell organischer Farbstoffe. Angew. Chem. 71, 93 (1959).
257. KUHN, L. P., R. E. LUTZ and C. R. BAUER: A Spectroscopic Study of cis- and trans-Dibenzoylethylenes and Related Compounds. J. Amer. Chem. Soc. 72, 5058 (1950).
258. KUHN, R.: Cis-trans-Umlagerungen der Äthylenkörper. In: K. FREUDENBERG, Stereochemie, S. 913. Leipzig und Wien: F. Deuticke. 1933.
258 a. — Molekulare Umlagerungen organischer Verbindungen. Österr. Chem.-Ztg. 62, 184 (1961).
259. KUHN, R. und D. BLUM: Über Kumulene, X. cis-trans-Isomerie bei Dinitrotetraphenyl-kumulenen. Chem. Ber. 92, 1483 (1959).
260. KUHN, R. und P. J. DRUMM: Umkehrbare Hydrierung und Dehydrierung bei Polyenen. Ber. dtsch. chem. Ges. 65, 1458 (1932).
261. KUHN, R., P. J. DRUMM, M. HOFFER und E. F. MÖLLER: Farbreaktionen und Autoxydation von Hydropolyen-carbonsäure-estern. Ber. dtsch. chem. Ges. 65, 1785 (1932).
262. KUHN, R. und H. FISCHER: Über Kumulene, XII. cis-Polyene durch Partialhydrierung von Kumulenen. Chem. Ber. 93, 2285 (1960).
263. KUHN, R., H. H. INHOFFEN, H. A. STAAB und W. OTTING: Vergleich des cis-Crocetin-dimethylesters aus Safran mit 8.8'-cis-Crocetin-dimethylester. Ber. dtsch. chem. Ges. 86, 965 (1953).
264. KUHN, R. und J. JAHN: Über Kumulene, V. Chem. Ber. 86, 759 (1953).
265. KUHN, R. und H. KRAUCH: Kumulene, VIII. Reduktion von Acetylen-, Diacetylen- und Triacetylen-Glykolen mit Zinn-(II)-chlorid; Kumulene mit nur zwei aromatischen Substituenten. Chem. Ber. 88, 309 (1955).
266. KUHN, R. und E. LEDERER: Iso-carotin (Über das Vitamin des Wachstums, III. Mitt.). Ber. dtsch. chem. Ges. 65, 637 (1932).
267. KUHN, R. und K. L. SCHOLLER: Über Kumulene, VI. cis-trans-Isomere Bis-[2-nitrodiphenylen]-butatriene. Chem. Ber. 87, 598 (1954).
268. KUHN, R. und K. WALLENFELS: Über 11-Phenyl-undeca-pentaenal und 15-Phenyl-pentadeca-heptaenal. Ber. dtsch. chem. Ges. 70, 1331 (1937).
269. KUHN, R. und A. WINTERSTEIN: Über konjugierte Doppelbindungen. I. Synthese von Diphenyl-poly-enen. II. Synthese von Biphenylen-poly-enen. Helv. Chim. Acta 11, 87, 116 (1928).
270. — — Die Dihydroverbindung der isomeren Bixine und die Elektronen-Konfiguration der Polyene (Über konjugierte Doppelbindungen, XXIII. Mitt.). Ber. dtsch. chem. Ges. 65, 646 (1932).
271. — — Über einen licht-empfindlichen Carotin-Farbstoff aus Safran. Ber. dtsch. chem. Ges. 66, 209 (1933).
272. — — Über die Konstitution des Pikro-crocins und seine Beziehung zu den Carotin-Farbstoffen des Safrans. Ber. dtsch. chem. Ges. 67, 344 (1934).
273. KUHN, R., A. WINTERSTEIN und E. LEDERER: Zur Kenntnis der Xanthophylle. Z. physiol. Chem. (Hoppe-Seyler) 197, 141 (1931).
274. KUHN, R., H. ZAHN und K. L. SCHOLLER: Über stereoisomere 2,2'-Diamino-bisdiphenylen-äthylene. Liebigs Ann. Chem. 582, 196 (1953).

275. KUHN, W. und R. LANDOLT: Über den Photodichroismus fester Carotinoid-Lösungen. II. Helv. Chim. Acta 34, 1929 (1951).

276. LAMBERTSEN, G. and O. R. BRAEKKAN: Studies on the Vitamin A Components of Fish Livers: Determination and Origin. In: G. POPJÁK and E. LE BRETON, Biochemical Problems of Lipids, p. 56. London: Butterworths Sci. Publ. 1956.

277. LANDOLT, R. und W. KUHN: Über den Photodichroismus fester Carotinoid-Lösungen. I. Helv. Chim. Acta 34, 1900 (1951).

278. LEDERER, E. and M. LEDERER: Chromatography. A Review of Principles and Applications. 2nd ed. New York: Elsevier Publ. Co. 1957.

279. LEHMAN, R. W., J. M. DIETERLE, W. T. FISHER and S. R. AMES: Isomerization of Vitamin A in Aqueous Multivitamin Drop Preparations. J. Amer. Pharm. Assoc. 49, 363 (1960).

280. LEROSEN, A. L.: Continuous Washing Apparatus for Solutions in Organic Solvents. Ind. Eng. Chem., Analyt. Ed. 14, 165 (1942).

281. — A Method for Standardization of Chromatographic Analysis. J. Amer. Chem. Soc. 64, 1905 (1942).

282. LEROSEN, A. L. and L. ZECHMEISTER: Prolycopene. J. Amer. Chem. Soc. 64, 1075 (1942).

283. — — The Carotenoid Pigments of the Fruit of *Celastrus scandens* L. Arch. Biochemistry 1, 17 (1942).

284. LEWIS, G. N. and M. CALVIN: The Color of Organic Substances. Chem. Rev. 25, 273 (1939).

285. LEWIS, G. N., T. T. MAGEL and D. LIPKIN: The Absorption and Re-emission of Light by *cis-* and *trans-*Stilbenes and the Efficiency of their Photochemical Isomerization. J. Amer. Chem. Soc. 62, 2973 (1940).

286. LINDLAR, H.: Ein neuer Katalysator für selective Hydrierungen. Helv. Chim. Acta 35, 446 (1952).

287. LINNER, E., C. H. EUGSTER und P. KARRER: Carotinoidsynthesen. XVIII. Synthese des 1,18-Di-β-naphtyl-3,7,12,16-tetramethyl-octadeca-nonaens. Helv. Chim. Acta 38, 1869 (1955).

288. LOEB, J. N., P. K. BROWN and G. WALD: Photochemical and Stereochemical Properties of Carotenoids at Low Temperatures. 2. *Cis-trans* Isomerism and Steric Hindrance. Nature (London) 184, 617 (1959).

289. LOWE, J. S. and R. A. MORTON: Some Aspects of Vitamin A Metabolism· Vitamins and Horm. 14, 97 (1956).

290. LUNDE, K.: Note on the Infrared Spectra of Pro-γ-carotene and Neo-γ-carotene P. Acta Chem. Scand. 13, 2154 (1959).

291. LUNDE, K. and L. ZECHMEISTER: A Study of the Infrared Spectra of Some Stereoisomeric Diphenylpolyenes. Acta Chem. Scand. 8, 1421 (1954).

292. — — *cis-trans* Isomeric 1,6-Diphenylhexatrienes. J. Amer. Chem. Soc. 76, 2308 (1954).

293. — — Infrared Spectra and *cis-trans* Configurations of Some Carotenoid Pigments. J. Amer. Chem. Soc. 77, 1647 (1955).

294. — — On the Infrared Spectrum of *cis-*Stilbene. Naturwiss. (in press).

295. LYTHGOE, R. J.: The Absorption Spectra of Visual Purple and of Indicator Yellow. J. Physiol. (London) 89, 331 (1937).

296. MACKINNEY, G.: Carotenoids. Annu. Rev. Biochem. 21, 473 (1952).

297. MACKINNEY, G. and J. A. JENKINS: Inheritance of Carotenoid Differences in *Lycopersicon esculentum* Strains. Proc. Nat. Acad. Sci. (USA) 35, 284 (1949).

298. — — Carotenoid Differences in Tomatoes. Proc. Nat. Acad. Sci. (USA) 38, 48 (1952).

299. MAGOON, E. F. and L. ZECHMEISTER: On the *cis* Forms of Some Biphenylene Derivatives of Butadiene and Hexatriene. J. Amer. Chem. Soc. 77, 5642 (1955).

300. — — On a *cis*-Neurosporene *ex Pyracantha* and the *in vitro* Stereoisomerization of Neurosporene. Arch. Biochem. Biophys. 68, 263 (1957).

301. — — Stepwise Stereoisomerization of Prolycopene, a Poly*cis* Carotenoid, to all-*trans*-Lycopene. Arch. Biochem. Biophys. 69, 535 (1957).

302. MAITLAND, P. and W. H. MILLS: Experimental Demonstration of the Allene Asymmetry. Nature (London) 135, 994 (1935).

303. — — Resolution of an Allene Hydrocarbon into Optical Antipodes by Asymmetric Catalysis. J. Chem. Soc. (London) 1936, 987.

304. MANUNTA, C.: The Manner of Action of the Genes Responsible for the Color of the Fruit in Cultivated Strains of Tomato *(Lycopersicon esculentum)* and in Wild Species of the Genus. Genet. agrar. (Pavia) 3, 38 pp. (1951) [Chem. Abstr. 48, 9476 (1954)].

305. MARTÍNEZ GARCÍA, F.: The Carotenoids of Red Pepper. Rev. fac. ciênc., Univ. Coimbra 20, 21 (1951) [Chem. Abstr. 46, 8287 (1952)].

306. — The Presence of Stereoisomerism of Capsorubin in the Pimiento. Farmacognosia (Madrid) 12, 169 (1952) [Chem. Abstr. 48, 5443 (1954)].

306 a. MARUSICH, W., E. DE RITTER, J. VREELAND and R. KRUKAR: Vitamin A Activity of Beta-apo-8′-carotenal. Agric. Food Chem. 8, 390 (1960).

307. MASE, Y., W. J. RABOURN and F. W. QUACKENBUSH: Carotene Production by *Penicillium sclerotiorum*. Arch. Biochem. Biophys. 68, 150 (1957).

308. MAYO, F. R. and C. WALLING: The Peroxide Effect in the Addition of Reagents to Unsaturated Compounds and in Rearrangement Reactions. Chem. Rev. 27, 351 (1940).

309. McCONNELL, H.: Catalysis of *Cis-Trans* Isomerization by Paramagnetic Substances. J. Chem. Physics 20, 1043 (1952).

310. McNICHOLAS, H. J.: The Visible and Ultraviolet Absorption Spectra of Carotene and Xanthophyll and the Changes Accompanying Oxydation. Bureau Standards J. Res. 7, 171 (1931).

311. MEUNIER, P. et J. JOUANNETEAU: Recherches sur l'isomérie *cis-trans* dans la série de la vitamine A (Axérophtol). Bull. soc. chim. biol. (Paris) 30, 260 (1948).

312. MEUNIER, P., J. JOUANNETEAU et G. ZWINGELSTEIN: Sur la coupure oxydante du β-carotène en rétinène (axérophtal) par MnO_2. C. R. hebd. Séances Acad. Sci. 231, 1170 (1950).

313. — — — Sur l'isomérisation *cis trans* des caroténoides en C_{40} provoquée par adsorption sur des oxydes métalliques. Bull. soc. chim. biol. (Paris) 33, 1228 (1951).

314. MIKHAÏLOV, B. M. and G. S. TER-SARKISYAN: Polyene Compounds. I. Synthesis of 1,5-Aryl-substituted Derivatives of 1,3,5-Hexatriene. Izvest. Akad. Nauk USSR, Otdel. Khim. Nauk 1956, 1079 [Chem. Abstr. 51, 5025 (1957)].

315. MILAS, N. A.: Synthesis of Biologically Active Vitamin A Substances. Science (Washington) 103, 581 (1946).

316. MILAS, N. A., P. DAVIS, I. BELIČ and D. A. FLEŠ: Synthesis of β-Carotene. J. Amer. Chem. Soc. 72, 4844 (1950).

317. MILAS, N. A., E. SAKAL, J. T. PLATI, J. T. RIVERS, J. K. GLADDING, F. X. GROSSI, Z. WEISS, M. A. CAMPBELL and H. F. WRIGHT: Synthesis of Products Related to Vitamin A. VI. The Synthesis of Biologically Active Vitamin A Ethers. J. Amer. Chem. Soc. 70, 1597 (1948).

318. MILLER, E. S.: A Precise Method, with Detailed Calibration for the Determination of Absorption Coefficients; the Quantitative Measurement of the Visible and Ultraviolet Absorption Spectra of α-Carotene, β-Carotene, and Lycopene. Plant Physiol. **12**, 667 (1937).

319. — Quantitative Biological Spectroscopy. Absorption Spectra, Vol. I. Minneapolis: Burgess Publ. Co. 1940.

320. MOORE, T.: Vitamin A and Carotene. I. The Association of Vitamin A Activity with Carotene in the Carrot Root. Biochemic. J. **23**, 803 (1929).

321. — Vitamin A. New York: Elsevier Publ. Co.; London: Cleaver-Hume. 1957.

322. MORTON, R. A.: Chemical Aspects of the Visual Process. Nature (London) **153**, 69 (1944).

323. MORTON, R. A. and F. BRO-RASMUSSEN: Comments on the Determination of Vitamin A in Natural Products and Especially in Cod-liver Oils. Analyst **80**, 410 (1955).

324. MORTON, R. A. and T. W. GOODWIN: Preparation of Retinene in vitro. Nature (London) **153**, 405 (1944).

325. — — Carotenoids and Vitamin A. Brit. Med. Bull. **12**, 37 (1956).

326. MORTON, R. A. and G. A. J. PITT: Studies on Rhodopsin. 9. pH and the Hydrolysis of Indicator Yellow. Biochemic. J. **59**, 128 (1955).

327. — — Visual Pigments. Fortschr. Chem. organ. Naturstoffe **14**, 244 (1957).

328. MORTON, R. A., M. K. SALAH and A. L. STUBBS: Retinene$_2$ and Vitamin A$_2$. Nature (London) **159**, 744 (1947).

329. MULLIKEN, R. S.: Intensities of Electronic Transitions in Molecular Spectra. VII. Conjugated Polyenes and Carotenoids. J. Chem. Physics **7**, 364 (1939).

330. — Structure and Ultraviolet Spectra of Ethylene, Butadiene and their Alkyl Derivatives. Rev. Modern Physics **14**, 265 (1942).

331. — Quantum-mechanical Methods and the Electronic Spectra and Structure of Molecules. Chem. Rev. **41**, 201 (1947).

332. NASH, H. A. and F. P. ZSCHEILE: The *cis-trans* Isomerization of α-Carotene Isomers. Arch. Biochemistry **5**, 77 (1944).

333. NAYLER, P. and M. C. WHITING: Researches on Polyenes. II. The Synthesis of Cosmene. J. Chem. Soc. (London) **1954**, 4006.

334. OROSHNIK, W.: Synthesis of Polyenes. V. α-Vitamin A Methyl Ether. J. Amer. Chem. Soc. **76**, 5499 (1954).

335. — The Synthesis and Configuration of Neo-b Vitamin A and Neoretinene b. J. Amer. Chem. Soc. **78**, 2651 (1956).

336. OROSHNIK, W., P. K. BROWN, R. HUBBARD and G. WALD: Hindered *cis* Isomers of Vitamin A and Retinene: the Structure of the Neo-b Isomer. Proc. Nat. Acad. Sci. (USA) **42**, 578 (1956).

337. OROSHNIK, W., G. KARMAS and A. D. MEBANE: Synthesis of Polyenes. I. *Retro*vitamin A Methyl Ether. Spectral Relationships between the β-Ionylidene and *Retro*ionylidene Series. J. Amer. Chem. Soc. **74**, 295 (1952).

338. — — — Synthesis of Polyenes. II. Allylic Rearrangement and Dehydrations in Substituted β-Ionols. J. Amer. Chem. Soc. **74**, 3807 (1952).

339. OROSHNIK, W. and A. D. MEBANE: Synthesis of Polyenes. VI. Isoprenoid Polyenes Containing Sterically Hindered *cis* Configurations. J. Amer. Chem. Soc. **76**, 5719 (1954).

340. OROSHNIK, W., A. D. MEBANE and G. KARMAS: Synthesis of Polyenes. III. Prototropic Rearrangements in β-Ionols and Related Compounds. J. Amer. Chem. Soc. **75**, 1050 (1953).

341. Ott, E. und R. Schröter: Über die Halbhydrierung der Acetylen-Bindung und die Abhängigkeit der geometrischen Konfiguration der entstehenden Äthylen-Verbindungen von der Reaktionsgeschwindigkeit. Ber. dtsch. chem. Ges. 60, 624 (1927).

342. Otting, W.: Über Kumulene, VII. Die Ultrarotspektren einiger Kumulene und Acetylenglykole. Chem. Ber. 87, 611 (1954).

343. Otto, R. und F. Stoffel: Das zweite Stilben. Ber. dtsch. chem. Ges. 30, 1799 (1897).

344. Paal, C. und W. Hartmann: Über katalytische Wirkungen kolloidaler Metalle der Platingruppe. VIII. Die stufenweise Reduktion der Phenyl-propiolsäure. Ber. dtsch. chem. Ges. 42, 3930 (1909).

344 a. Para, J. and L. S. Forster: Thermal *cis-trans* Isomerization of Diphenylhexatriene in Solution. Trans. Faraday Soc. 57, 87 (1961).

345. Paschke, R. F., W. Tolberg and D. H. Wheeler: *Cis,Trans* Isomerism of the Eleostearate Isomers. J. Amer. Oil Chem. Soc. 30, 97 (1953).

346. Paul, M. F., V. R. Ells and H. E. Paul: The Effect of Mineral Oil on Food Utilization. II. Changes of β-Carotene in Mineral Oil. Amer. J. Digestive Diseases 18, 278 (1951).

347. Pauling, L.: Recent Work on the Configuration and Electronic Structure of Molecules; with some Applications to Natural Products. Fortschr. Chem. organ. Naturstoffe 3, 203 (1939).

348. — A Theory of the Color of Dyes. Proc. Nat. Acad. Sci. (USA) 25, 577 (1939).

349. — Zur *cis-trans*-Isomerisierung von Carotinoiden. Helv. Chim. Acta 32, 2241 (1949).

350. Petracek, F. J. and L. Zechmeister: Stereoisomeric Phytofluenes. J. Amer. Chem. Soc. 74, 184 (1952).

351. — — Reaction of β-Carotene with N-Bromosuccinimide: The Formation and Conversions of Some Polyene Ketones. J. Amer. Chem. Soc. 78, 1427 (1956).

352. — — The Hydrolytic Cleavage Products of Boron Trifluoride Complexes of β-Carotene, Some Dehydrogenated Carotenes and Anhydrovitamin A_1. J. Amer. Chem. Soc. 78, 3188 (1956).

352 a. Petzold, E. N., F. W. Quackenbush and M. McQuistan: Zeacarotenes, New Provitamins A from Corn. Arch. Biochem. Biophys. 82, 117 (1959).

353. Pinckard, J. H.: Thesis, California Institute of Technology, Pasadena, 1949.

354. Pinckard, J. H., J. S. Kittredge, D. L. Fox, F. T. Haxo and L. Zechmeister: Pigments from a Marine "Red Water" Population of the Dinoflagellate *Prorocentrum micans*. Arch. Biochem. Biophys. 44, 189 (1953).

355. Pinckard, J. H., B. Wille and L. Zechmeister: A Comparative Study of the Three Stereoisomeric 1,4-Diphenylbutadienes. J. Amer. Chem. Soc. 70, 1938 (1948).

356. Pitt, G. A. J. and R. A. Morton: *cis-trans* Isomers of Retinene in Visual Processes. Biochemic. J. 72, 40 P (1959).

357. — — *cis-trans* Isomers of Retinene in Visual Processes. Biochem. Soc. Symposia (Cambridge), No. 19, 67 (1960).

358. Plack, P. A.: Maleic Anhydride in the Study of Naturally Occurring Isomers of Vitamin A. Biochemic. J. 64, 56 (1956).

359. — The Conversion of 11-*cis* into all-*trans* Vitamin A in the Rat. Brit. J. Nutrit. 13, 111 (1959).

360. — Vitamin A_1 Aldehyde in Hen's Eggs. Nature (London) 186, 234 (1960).

361. Plack, P. A., L. R. Fisher, K. M. Henry and S. K. Kon: *cis*-Isomers of Vitamin A in some Marine Crustacea. Biochemic. J. 64, 17 P (1956).

362. PLACK, P.·A., S. K. KON and S. Y. THOMPSON: Vitamin A_1 Aldehyde in the Eggs of the Herring (*Clupea harengus* L.) and other Marine Teleosts. Biochemic. J. **71**, 467 (1959).

363. PLATT, J. R.: Electronic Structure and Excitation of Polyenes and Porphyrins. In: A. HOLLAENDER, Radiation Biology, Vol. III, p. 71. New York: McGraw-Hill. 1956.

364. POLGÁR, A., C. B. VAN NIEL and L. ZECHMEISTER: Studies on the Pigments of the Purple Bacteria. II. A Spectroscopic and Stereochemical Investigation of Spirilloxanthin. Arch. Biochemistry **5**, 243 (1944).

365. POLGÁR, A. and L. ZECHMEISTER: Isomerization of β-Carotene. Isolation of a Stereoisomer with Increased Adsorption Affinity. J. Amer. Chem. Soc. **64**, 1856 (1942).

366. — — Action of Cold Concentrated Hydriodic Acid on Carotenes: Structure and *cis-trans* Isomerization of Some Reaction Products. J. Amer. Chem. Soc. **65**, 1528 (1943).

367. — — A Spectroscopic Study in the Stereoisomeric Capsanthin Set. *Cis*-peak Effect and Configuration. J. Amer. Chem. Soc. **66**, 186 (1944).

368. PORTER, J. W.: Relationships Between Physical Properties and Structure of Carotenes and Colorless Polyenes. Arch. Biochem. Biophys. **45**, 291 (1953).

369. PORTER, J. W. and R. E. LINCOLN: I. *Lycopersicon* Selections Containing a High Content of Carotenes and Colorless Polyenes. II. The Mechanism of Carotene Biosynthesis. Arch. Biochemistry **27**, 390 (1950).

370. PORTER, J. W. and F. P. ZSCHEILE: Carotenes of *Lycopersicon* Species and Strains. Arch. Biochemistry **10**, 537 (1946).

371. POUTET, J. J.: Procédé pour reconnaître la falsification de l'huile d'olive par celle de graines. Ann. chim. phys. [2] **12**, 58 (1819).

372. PRADHAN, S. K. and N. G. MAGAR: Vitamins A_1, A_2, and Neovitamin A in Shark-liver Oils. Indian J. Med. Res. **44**, 11 (1956) [Chem. Abstr. **50**, 9687 (1956)].

373. PRICE, C. C. and G. BERTI: The Polymerization of Stilbene in Boron Fluoride Etherate. J. Amer. Chem. Soc. **76**, 1219 (1954).

374. PRICE, C. C. and M. MEISTER: *cis-trans*-Isomerization with Boron Fluoride. J. Amer. Chem. Soc. **61**, 1595 (1939).

375. PULLMAN, A.: Caractéristiques électroniques des polyènes conjugués d'intérêt biologique (caroténoïdes, vitamines A, rétinènes). C. R. hebd. Séances Acad. Sci. **251**, 1430 (1960).

376. PULLMAN, A. and B. PULLMAN: The *cis-trans* Isomerization of Conjugated Polyenes and the Occurrence of a Hindered *cis*-Isomer of Retinene in the Rhodopsin System. Proc. Nat. Acad. Sci. (USA) **47**, 7 (1961).

377. QUACKENBUSH, F. W., H. STEENBOCK and W. H. PETERSON: The Effect of Acids on Carotenoids. J. Amer. Chem. Soc. **60**, 2937 (1938).

378. RABOURN, W. J. and F. W. QUACKENBUSH: The Occurrence of Phytoene in Various Plant Materials Arch. Biochem. Biophys. **44**, 159 (1953).

379. — — The Structure of Phytoene. Arch. Biochem. Biophys. **61**, 111 (1956).

380. — — A Unified Concept of Carotene Photosynthesis. Amer. Chem. Soc. Meeting, Sept. 1957. Abstr. p. 88 C.

381. RABOURN, W. J., F. W. QUACKENBUSH and J. W. PORTER: Isolation and Properties of Phytoene. Arch. Biochem. Biophys. **48**, 267 (1954).

382. RADDING, C. M. and G. WALD: The Action of Enzymes on Rhodopsin. J. Gen. Physiol. **42**, 371 (1958).

383. RASMUSSEN, R. S.: Infrared Spectroscopy in Structure Determination and its Application to Penicillin. Fortschr. Chem. organ. Naturstoffe **5**, 331 (1948).

384. RASMUSSEN, R. S. and R. R. BRATTAIN: Infra-red Absorption Spectra of the C_2 to C_4 Mono-olefins and of 2-Methyl-2-butene. J. Chem. Physics **15**, 120 (1947).

385. — — Infra-red Absorption Spectra of Some C_4 and C_5 Dienes. J. Chem. Physics **15**, 131 (1947).

386. — — Infrared Spectra of Some Carboxylic Acid Derivatives. J. Amer. Chem. Soc. **71**, 1073 (1949).

387. RAU, W. und C. ZEHENDER: Die Carotinoide von *Fusarium aquaeductum* LAGH. Arch. Mikrobiol. **32**, 423 (1959).

387 a. RIEZEBOS, G. and E. HAVINGA: The Influence of Non Planarity in Styrene and Stilbene Derivatives, III. Syntheses and Ultra-violet Spectra of 4-4′-Substituted *cis-* and *trans*-Stilbenes. Rec. trav. chim. Pays-Bas **80**, 446 (1961).

388. ROBESON, C. D. and J. G. BAXTER: A new Vitamin A. Nature (London) **155**, 300 (1945).

389. — — Neovitamin A. J. Amer. Chem. Soc. **69**, 136 (1947).

390. ROBESON, C. D., W. P. BLUM, J. M. DIETERLE, J. D. CAWLEY and J. G. BAXTER: Chemistry of Vitamin A. XXV. Geometrical Isomers of Vitamin A Aldehyde and an Isomer of its α-Ionone Analog. J. Amer. Chem. Soc. **77**, 4120 (1955).

391. ROBESON, C. D., J. D. CAWLEY, L. WEISLER, M. H. STERN, C. C. EDDINGER and A. J. CHECHAK: Chemistry of Vitamin A. XXIV. The Synthesis of Geometric Isomers of Vitamin A *via* Methyl β-Methylglutaconate. J. Amer. Chem. Soc. **77**, 4111 (1955).

392. ROSENBERG, B.: Photoconduction and *Cis-Trans* Isomerism in β-Carotene. J. Chem. Physics **31**, 238 (1959); see also **34**, 63 (1961).

393. RÜEGG, R., H. LINDLAR, M. MONTAVON, G. SAUCY, S. F. SCHAEREN, U. SCHWIETER und O. ISLER: Synthesen in der Carotinoid-Reihe. 14. Mitt. Synthese von β-Apo-12′-carotinal (C_{25}). Helv. Chim. Acta **42**, 847 (1959).

394. RÜEGG, R., M. MONTAVON, G. RYSER, G. SAUCY, U. SCHWIETER und O. ISLER: Synthesen in der Carotinoid-Reihe. 15. Mitt. Synthesen in der β-Carotinal- und β-Carotinol-Reihe. Helv. Chim. Acta **42**, 854 (1959).

394a. RÜEGG, R., U. SCHWIETER, G. RYSER, P. SCHUDEL und O. ISLER: Synthesen in der Carotinoid-Reihe. 17. Mitt. γ-Carotin sowie *d,l*-α- und β-Carotin aus Dehydro-β-apo-12′-carotinal (C_{25}). — 18. Mitt. Synthese von 7′,8′-Dihydro-γ-carotin (β-Zeacarotin) und 3′,4′-Dehydro-γ-carotin (Torulin). Helv. Chim. Acta **44**, 985, 994 (1961).

395. SANDOVAL, A. and L. ZECHMEISTER: Some Spectroscopic Changes Connected with the Stereoisomerization of Diphenylbutadiene. J. Amer. Chem. Soc. **69**, 553 (1947).

396. SAPERSTEIN, S. and M. P. STARR: Association of Carotenoid Pigments with Protein Components in non-photosynthetic Bacteria. Biochim. Biophys. Acta **16**, 482 (1955).

397. SAVINOV, B. G. and A. A. MIKHAÏLOVNINA: Neo-β-carotene as the Product of Primary Stereoisomeric Transformation of β-Carotene on Heating. Dokl. Akad. Nauk USSR **88**, 887 (1953) [Chem. Abstr. **48**, 3311 (1954)].

398. SCHROEDER, W. A.: Formation of Pro-carotenoids in "Monkey Flowers" under Some Conditions. J. Amer. Chem. Soc. **64**, 2510 (1942).

399. SCHULTE-FROHLINDE, D.: Quantenausbeuten der photochemischen *cis* ⇄ *trans*-Umlagerung von p-Nitro-p′-methoxy-stilben in verschiedenen Lösungsmitteln. Liebigs Ann. Chem. **615**, 114 (1958).

400. SEITZ, K., Hs. H. GÜNTHARD und O. JEGER: Veilchenriechstoffe. 37. Mitt. Über die Trennung von α- und β-Jonon durch fraktionierte Destillation. Helv. Chim. Acta **33**, 2196 (1950).

401. SHANTZ, E. M.: Rehydro Vitamin A, the Compound from Anhydro Vitamin A in vivo. J. Biol. Chem. **182**, 515 (1950).

402. SHANTZ, E. M., J. D. CAWLEY and N. D. EMBREE: Anhydro ("Cyclized") Vitamin A. J. Amer. Chem. Soc. **65**, 901 (1943).

403. SHECHTER, H., J. J. GARDIKES and A. H. PAGANO: Stereochemistry of Addition of Dinitrogen Tetroxide to *cis* and *trans* Stilbenes. J. Amer. Chem. Soc. **81**, 5420 (1959).

404. SHEPPARD, N. and D. M. SIMPSON: The Infra-red and Raman Spectra of Hydrocarbons. I. Acetylenes and Olefins. Quart. Rev. Chem. Soc. (London) **6**, 1 (1952).

405. SHEPPARD, N. and G. B. B. M. SUTHERLAND: Vibration Spectra of Hydrocarbon Molecules. I. Frequencies due to Deformation Vibrations of Hydrogen Atoms Attached to a Double Bond. Proc. Roy. Soc. (London), Ser. A **196**, 195 (1949).

406. SIMPSON, W. T.: Resonance Force Theory of Carotenoid Pigments. J. Amer. Chem. Soc. **77**, 6164 (1955).

407. SLY, W. G.: A Preliminary Report on the Crystal-structure Determination of 15,15′-Dehydro-β-carotene. Acta Crystallogr. **8**, 115 (1955). — Thesis, California Institute of Technology. 1955.

408. SMAKULA, A.: Über die photochemische Umwandlung des *trans*-Stilbens. Z. physik. Chem. B **25**, 90 (1934).

409. STAINER, D. W. and T. K. MURRAY: Isomerization of Vitamin A by Tissue Homogenates. Canad. J. Biochem. Physiol. **38**, 1467 (1960).

410. STAINER, D. W., T. K. MURRAY and J. A. CAMPBELL: Isomerization of 11-*cis*-Vitamin A in vivo. Canad. J. Biochem. Physiol. **38**, 1219 (1960).

411. STARR, M. P. and S. SAPERSTEIN: Thiamine and the Carotenoid Pigments of *Corynebacterium poinsettiae*. Arch. Biochem. Biophys. **43**, 157 (1953).

412. STEENBOCK, H., M. T. SELL, E. M. NELSON and M. V. BUELL: The Fat-soluble Vitamine. J. Biol. Chem. **46**, XXXII (1921).

413. STEGEMEYER, H.: Die Infrarot-Absorption des *cis*-Stilbens im Valenz-schwingungsgebiet. Naturwiss. **48**, 128 (1961).

414. STITT, F., E. M. BICKOFF, G. F. BAILEY, C. R. THOMPSON and S. FRIEDLANDER: Spectrophotometric Determination of β-Carotene Stereoisomers in Alfalfa. J. Assoc. Off. Agric. Chemists **34**, 460 (1951).

415. STOERMER, R.: Über die Umlagerung stabiler stereoisomerer Äthylenkörper in labile durch ultraviolettes Licht (I). Ber. dtsch. chem. Ges. **42**, 4865 (1909).

416. STRAIN, H. H.: Leaf Xanthophylls. Carnegie Inst. Washington Publ. No. 490. Washington. 1938.

417. — Carotene. XI. Isolation and Detection of α-Carotene, and the Carotenes of Carrot Roots and of Butter. J. Biol. Chem. **127**, 191 (1939).

418. — Isomerizations of Polyene Acids and Carotenoids. Preparation of β-Eleo-stearic and β-Licanic Acids. J. Amer. Chem. Soc. **63**, 3448 (1941).

419. — Chromatographic Adsorption Analysis. New York: Interscience Publ. 1942.

420. — Problems in Chromatography and in Colloid Chemistry Illustrated by Leaf Pigments. J. Physic. Chem. **46**, 1151 (1942).

421. — Chloroplast Pigments. Annu. Rev. Biochem. **13**, 591 (1944).

422. — Molecular Structure and Adsorption Sequences of Carotenoid Pigments. J. Amer. Chem. Soc. **70**, 588 (1948).

423. — Leaf Xanthophylls: The Action of Acids on Violaxanthin, Violeoxanthin, Taraxanthin and Tareoxanthin. Arch. Biochem. Biophys. **48**, 458 (1954).

424. STRAIN, H. H. and W. M. MANNING: The Occurrence and Interconversion of Various Fucoxanthins. J. Amer. Chem. Soc. **64**, 1235 (1942).

425. STRAIN, H. H. and W. M. MANNING: A Unique Polyene Pigment of the Marine Diatom *Navicula Torquatum.* J. Amer. Chem. Soc. 65, 2258 (1943).

426. STRAIN, H. H., W. M. MANNING and G. HARDIN: Xanthophylls and Carotenes of Diatoms, Brown Algae, Dinoflagellates, and Sea-anemones. Biol. Bull. 86, 169 (1944).

427. STRAUS, F.: Zur Kenntnis der Acetylenbindung. Liebigs Ann. Chem. 342, 190 (1905).

428. SURMATIS, J. D., J. MARICQ and A. OFNER: Synthesis of Carotene Homologs. J. Organ. Chem. (USA) 23, 157 (1958).

429. SUZUKI, N. and K. TSUKIDA: Carotenoids of the Flowers of *Osmanthus fragrans.* Chem. pharm. Bull. (Tokyo) 7, 133 (1959).

430. — — On Some *cis*-Forms of Luteochrome. Chem. pharm. Bull. (Tokyo) 7, 878 (1959).

431. SZASZ, G. J. and N. SHEPPARD: An Infra-red Spectroscopic Study of the Configuration of Some Chlorinated Butadienes. Trans. Faraday Soc. 49, 358 (1953).

432. TAPPI, G. und P. KARRER: Über die Carotinoide aus den Staubbeuteln von *Lilium candidum.* *Cis*-Antheraxanthin. Helv. Chim. Acta 32, 50 (1949).

433. TAPPI, G. e E. MENZIANI: Sui pigmenti del polline di *Lilium mantchiuricum.* Atti Soc. Nat. Mat. Modena 85, 28 (1954).

434. TAYLOR, T. W. J. and C. E. J. CRAWFORD: Preparation of *iso*Stilbene. J. Chem. Soc. (London) 1934, 1130.

435. TAYLOR, T. W. J. and A. R. MURRAY: Isomeric Change in Certain Stilbenes. J. Chem. Soc. (London) 1938, 2078.

436. THOMAS, J. F. and G. BRANCH: The Principal Electronic Absorption Bands of the Vinylogous Series Derived from Benzylaldehyde and Benzophenone. J. Amer. Chem. Soc. 75, 4793 (1953).

437. THOMPSON, C. R., E. M. BICKOFF and W. D. MACLAY: Formation of Stereoisomers of β-Carotene in Alfalfa. Ind. Eng. Chem. 43, 126 (1951).

438. TOMES, M. L., F. W. QUACKENBUSH, O. E. NELSON, Jr. and B. NORTH: The Inheritance of Carotenoid Pigment Systems in the Tomato. Genetics 38, 117 (1953).

439. TROMBLY, H. H. and J. W. PORTER: Additional Carotenes and a Colorless Polyene of *Lycopersicon* Species and Strains. Arch. Biochem. Biophys. 43, 443 (1953).

440. TSUKIDA, K. and L. ZECHMEISTER: The Stereoisomerization of β-Carotene Epoxides and the Simultaneous Formation of Furanoid Oxides. Arch. Biochem. Biophys. 74, 408 (1958).

441. URUSHIBARA, Y. and O. SIMAMURA: The Effect of Oxygen and Reduced Nickel on the Catalytic Action of Hydrogen Bromide in the Isomerization of Isostilbene into Stilbene. Bull. Chem. Soc. Japan 12, 507 (1937).

442. — — Isomerization of Isostilbene to Stilbene by Hydrogen Bromide in the Presence of Oxygen and of Ferromagnetic Metals. Bull. Chem. Soc. Japan 13, 566 (1938).

443. VALLENTYNE, J. R.: Carotenoids in a 20,000-Year-Old Sediment from Searles Lake, California. Arch. Biochem. Biophys. 70, 29 (1957).

444. VAN'T HOFF, H. J.: Sur les formules de structure dans l'espace. Bull. soc. chim. (Paris) [2] 23, 295 (1875).

445. — La chimie dans l'espace. Rotterdam: P. M. Bazendijk. 1875.

446. WAGNER, R., J. FINE, J. W. SIMMONS and J. H. GOLDSTEIN: Microwave Spectrum, Structure, and Dipole Moment of *s-trans* Acrolein. J. Chem. Physics 26, 634 (1957).

447. WAIGHT, E. S. and R. L. ERSKINE: Absorption Spectra of Conjugated Carbonyl Compounds. In: G. W. GRAY, Steric Effects in Conjugated Systems (Chem. Soc. Symposium), p. 73. London: Butterworths. 1958.

448. WALD, G.: Vitamin A in the Retina. Nature (London) **132**, 316 (1933).

449. — Carotenoids and the Vitamin A Cycle in Vision. Nature (London) **134**, 65 (1934).

450. — Vitamin A in Eye Tissues. J. Gen. Physiol. **18**, 905 (1934).

451. — Carotenoids and the Visual Cycle. J. Gen. Physiol. **19**, 351 (1935).

452. — Pigments of the Retina. I. The Bull Frog. J. Gen. Physiol. **19**, 781 (1936).

453. — Visual Purple System in Fresh-water Fishes. Nature (London) **139**, 1017 (1937).

454. — The Porphyropsin Visual System. J. Gen. Physiol. **22**, 775 (1939).

455. — The Synthesis from Vitamin A_1 of "Retinene$_1$," and of a new 545 mμ Chromogen yielding Light-sensitive Products. J. Gen. Physiol. **31**, 489 (1948).

456. — The Enzymatic Reduction of the Retinenes to the Vitamins A. Science (Washington) **109**, 482 (1949).

457. — The Biochemistry of Vitamin A. In: R. M. HERRIOTT, Symposium on Nutrition. Baltimore: Johns Hopkins Press. 1953.

458. — The Biochemistry of Vision. Annu. Rev. Biochem. **22**, 497 (1953).

459. — Photochemical and Stereochemical Properties of Carotenoids at Low Temperatures. 3. Discussion. Nature (London) **184**, 620 (1959).

460. — The Visual Function of the Vitamins A. Vitamins and Horm. **18**, 417 (1960).

461. — The Molecular Organization of Visual Systems. In: Light and Life. Baltimore: Johns Hopkins Press. 1961.

462. — General Discussion of Retinal Structure in Relation to the Visual Process. In: The Structure of the Eye. New York: Academic Press. 1961.

463. WALD, G. and P. K. BROWN: The Synthesis of Rhodopsin from Retinene$_1$. Proc. Nat. Acad. Sci. (USA) **36**, 84 (1950).

464. — — Synthesis and Bleaching of Rhodopsin. Nature (London) **177**, 174 (1956).

465. — — The Vitamin A of a Euphausiid Crustacean. J. Gen. Physiol. **40**, 627 (1957).

466. — — Human Rhodopsin. Science (Washington) **127**, 222 (1958).

467. WALD, G., P. K. BROWN, R. HUBBARD and W. OROSHNIK: Hindered *cis* Isomers of Vitamin A and Retinene: The Structure of the neo-b Isomer. Proc. Nat. Acad. Sci. (USA) **41**, 438 (1955).

468. WALD, G., P. K. BROWN and P. H. SMITH: Cyanopsin, a New Pigment of Cone Vision. Science **118**, 505 (1953).

469. — — — Iodopsin. J. Gen. Physiol. **38**, 623 (1955).

470. WALD, G. and S. P. BURG: Crustacean Vitamin A. Federat. Proc. (Amer. Soc. exp. Biol.) **14**, 300 (1955).

471. WALD, G., J. DURELL and R. C. C. ST. GEORGE: The Light Reaction in the Bleaching of Rhodopsin. Science (Washington) **111**, 179 (1950).

472. WALD, G. and R. HUBBARD: Reduction of Retinene$_1$ to Vitamin A_1 in vitro. J. Gen. Physiol. **32**, 367 (1949).

473. — — Enzymic Aspects of the Visual Processes. In: The Enzymes, Vol. 3, p. 369, 2nd Ed. New York: Academic Press. 1960.

474. WALLACE, V. and J. W. PORTER: Phytofluene. Arch. Biochem. Biophys. **36**, 468 (1952).

475. WALLCAVE, L.: Thesis, California Institute of Technology, Pasadena, 1953.

476. WALLCAVE, L., J. LEEMANN and L. ZECHMEISTER: Action of Boron Trifluoride Etherate on β-Carotene. Proc. Nat. Acad. Sci. (USA) **39**, 604 (1953).

477. WALLCAVE, L. and L. ZECHMEISTER: Conversion of Dehydro-β-carotene, via its Boron Trifluoride Complex, into an Isomer of Cryptoxanthin. J. Amer. Chem. Soc. 75, 4495 (1953).

478. WARREN, C. K. and B. C. L. WEEDON: Carotenoids and Related Compounds. VI. Some Conjugated Polyene Diketones, and their Comparison with Capsorubin. J. Chem. Soc. (London) 1958, 3972.

479. WENDLER, N. L., C. ROSENBLUM and M. TISHLER: The Oxidation of β-Carotene. J. Amer. Chem. Soc. 72, 234 (1950).

480. WILLSTÄTTER, R. und A. STOLL: Untersuchungen über Chlorophyll. Methoden und Ergebnisse. Berlin: Julius Springer. 1913.

481. WINTERSTEIN, A. und B. HEGEDÜS: Über das Vorkommen des Retinens in der Natur. Z. physiol. Chem. (Hoppe-Seyler) 321, 97 (1960).

481 a. WINTERSTEIN, A., A. STUDER und R. RÜEGG: Neuere Ergebnisse der Carotinoidforschung. Chem. Ber. 93, 2951 (1960).

482. WISEMAN, H. G., S. S. STONE, H. L. SAVAGE and L. A. MOORE: Action of Celites on Carotene and Lutein. Analyt. Chemistry 24, 681 (1952).

483. WITTIG, G. und U. SCHÖLLKOPF: Über Triphenyl-phosphin-methylene als olefinbildende Reagenzien. I. Mitt. Chem. Ber. 87, 1318 (1954).

484. WITTIG, G. und W. WIEMER: Zur Valenztautomerie ungesättigter Systeme. Liebigs Ann. Chem. 483, 144 (1930).

485. WÜRSCH, J. und U. SCHWIETER: Synthese von β-Carotin-[6,6'-^{14}C]. Helv. Chim. Acta 39, 1067 (1956).

486. WYMAN, G. M.: The cis-trans Isomerization of Conjugated Compounds. Chem. Rev. 55, 625 (1955).

487. YAMAGUCHI, M.: Chemical Constitution of Renieratene. Bull. Chem. Soc. Japan 30, 979 (1957).

488. — Chemical Constitution of Isorenieratene. Bull. Chem. Soc. Japan 31, 51 (1958).

489. — Renieratene, a New Carotenoid Containing Benzene Rings, Isolated from a Sea Sponge. Bull. Chem. Soc. Japan 31, 739 (1958).

490. ZALOKAR, M.: Isolation of an Acidic Pigment in Neurospora. Arch. Biochem. Biophys. 70, 568 (1957).

491. ZECHMEISTER, L.: Carotinoide. Ein biochemischer Bericht über pflanzliche und tierische Polyenfarbstoffe. Berlin: Julius Springer. 1934.

492. — cis-trans Isomerization and Stereochemistry of Carotenoids and Diphenylpolyenes. Chem. Rev. 34, 267 (1944).

493. — Stereochemistry and Chromatography. Ann. New York Acad. Sci. 49, 220 (1948).

494. — Adsorption and Some Constitutional and Steric Properties. Discuss. Faraday Soc. 7, 54 (1949).

495. — Stereoisomeric Provitamins A. Vitamins and Horm. 7, 57 (1949).

496. — Les provitamines A stéréoïsomériques. Bull. Soc. chim. biol. (Paris) 31, 956 (1949).

497. — Progress in Chromatography 1938–1947. London: Chapman & Hall; New York: J. Wiley. 1950.

498. — Some Stereochemical Aspects of Polyenes. Experientia 10, 1 (1954).

499. — Some in vitro Conversions of Naturally Occurring Carotenoids. Fortschr. Chem. organ. Naturstoffe 15, 31 (1958).

500. — Bibliography of Papers Published by L. Zechmeister and Co-authors in the Fields of Chemistry and Biochemistry 1913–1958. Wien: Springer-Verlag. 1958.

501. ZECHMEISTER, L. und L. v. CHOLNOKY: Die chromatographische Adsorptionsmethode. Wien: Julius Springer. 1937.

502. ZECHMEISTER, L. und L. v. CHOLNOKY: Untersuchungen über den Paprika-Farbstoff. X. Citraurin aus Capsanthin. Liebigs Ann. Chem. **530**, 291 (1937).

503. — — Untersuchungen über den Paprika-Farbstoff. XI. Isomerisierungs-Erscheinungen. Liebigs Ann. Chem. **543**, 248 (1940).

504. ZECHMEISTER, L., L. v. CHOLNOKY und A. POLGÁR: Über die Isomerisierung des Zeaxanthins und Physaliens. Ber. dtsch. chem. Ges. **72**, 1678 (1939).

505. — — — Zur Isomerisierung von Xanthophyllen (Nachtrag). Ber. dtsch. chem. Ges. **72**, 2039 (1939).

506. ZECHMEISTER, L., H. J. DEUEL, Jr., H. H. INHOFFEN, J. LEEMANN, S. M. GREENBERG and J. GANGULY: Stereochemical Configuration and Provitamin A Activity. X. A Comparison of Synthetic 15,15′-Mono*cis*-β-carotene (Central Mono*cis*-β-carotene) with All-*trans*-β-carotene in the Rat and Chick. Arch. Biochem. Biophys. **36**, 80 (1952).

507. ZECHMEISTER, L. and R. B. ESCUE: New Stereoisomers of Methylbixin. Science (Washington) **96**, 229 (1942).

508. — — Isolation of Prolycopene and Pro-γ-carotene from *Evonymus fortunei*. J. Biol. Chem. **144**, 321 (1942).

509. — — A Stereochemical Study of Methylbixin. J. Amer. Chem. Soc. **66**, 322 (1944).

510. ZECHMEISTER, L. and G. KARMAKAR: The Occurrence of Phytofluene in Green Plant Organs. Arch. Biochem. Biophys. **47**, 160 (1953).

511. ZECHMEISTER, L. and B. K. KOE: Stepwise Dehydrogenation of the Colorless Polyenes Phytoene and Phytofluene with N-Bromosuccinimide to Carotenoid Pigments. J. Amer. Chem. Soc. **76**, 2923 (1954).

512. ZECHMEISTER, L. and R. M. LEMMON: Contribution to the Stereochemistry of Cryptoxanthin and Zeaxanthin. J. Amer. Chem. Soc. **66**, 317 (1944).

513. ZECHMEISTER, L. and A. L. LeROSEN: Contribution to the Stereochemistry of Diphenylpolyenes. Science (Washington) **95**, 587 (1942).

514. — — Stereoisomeric Diphenyloctatetraenes. J. Amer. Chem. Soc. **64**, 2755 (1942).

515. ZECHMEISTER, L., A. L. LeROSEN, W. A. SCHROEDER, A. POLGÁR and L. PAULING: Spectral Characteristics and Configuration of Some Stereoisomeric Carotenoids Including Prolycopene and Pro-γ-carotene. J. Amer. Chem. Soc. **65**, 1940 (1943).

516. ZECHMEISTER, L., A. L. LeROSEN, F. W. WENT and L. PAULING: Prolycopene, a Naturally Occurring Stereoisomer of Lycopene. Proc. Nat. Acad. Sci. (USA) **27**, 468 (1941).

517. ZECHMEISTER, L. and E. F. MAGOON: Spectral Maxima of Stereoisomeric Polyenes. Chem. and Ind. **1957**, 431.

518. ZECHMEISTER, L. and W. H. McNEELY: Separation of *cis* and *trans* Stilbenes by Application of the Chromatographic Brush Method. J. Amer. Chem. Soc. **64**, 1919 (1942).

519. ZECHMEISTER, L. and F. J. PETRACEK: Absence of Detectable Poly-*cis* Forms from Heat-isomerized Lycopene Solutions. J. Amer. Chem. Soc. **74**, 282 (1952).

520. — — On the Structure of the Deoxyluteins. Arch. Biochem. Biophys. **61**, 243 (1956).

521. ZECHMEISTER, L. and J. H. PINCKARD: Some Poly*cis*-lycopenes Occurring in the Fruit of *Pyracantha*. J. Amer. Chem. Soc. **69**, 1930 (1947).

522. — — Stereoisomeric Diphenyloctatetraenes. II. J. Amer. Chem. Soc. **76**, 4144 (1954).

523. ZECHMEISTER, L., J. H. PINCKARD, S. M. GREENBERG, E. STRAUB, T. FUKUI, and H. J. DEUEL, Jr.: Stereochemical Configuration and Provitamin A Activity. VIII. Pro-γ-carotene (a Poly-*cis* Compound) and its All-*trans* Isomer in the Rat. Arch. Biochemistry **23**, 242 (1949).

524. ZECHMEISTER, L. and A. POLGÁR: *cis-trans* Isomerization and Spectral Characteristics of Carotenoids and Some Related Compounds. J. Amer. Chem. Soc. **65**, 1522 (1943).

525. — — *cis-trans* Isomerization and *cis*-Peak Effect in the α-Carotene Set and in Some Other Stereoisomeric Sets. J. Amer. Chem. Soc. **66**, 137 (1944).

526. — — Contributions to the Stereochemistry of γ-Carotene. J. Amer. Chem. Soc. **67**, 108 (1945).

527. ZECHMEISTER, L. and A. SANDOVAL: The Occurrence and Estimation of Phytofluene in Plants. Arch. Biochemistry **8**, 425 (1945).

528. ZECHMEISTER, L. and W. A. SCHROEDER: On the Occurrence of Stereoisomeric Carotenoids in Nature. Science (Washington) **94**, 609 (1941).

529. — — The Pigment of *Mimulus longiflorus* and the Isolation of its γ-Carotene Component. Arch. Biochemistry **1**, 231 (1942).

530. — — The Fruit of *Pyracantha angustifolia*: a Practical Source of Pro-γ-carotene and Prolycopene. J. Biol. Chem. **144**, 315 (1942).

531. — — Pro-γ-carotene. J. Amer. Chem. Soc. **64**, 1173 (1942).

532. — — *cis-trans* Isomerization and Spectral Characteristics of Gazaniaxanthin. Further Evidence of its Structure. J. Amer. Chem. Soc. **65**, 1535 (1943).

533. ZECHMEISTER, L. und P. TUZSON: Über das Polyen-Pigment der Orange, II. Mitt. Citraurin. Ber. dtsch. chem. Ges. **70**, 1966 (1937).

534. — — Spontaneous Isomerization of Lycopene. Nature (London) **141**, 249 (1938).

535. — — Isomerization of Carotenoids. Biochemic. J. **32**, 1305 (1938).

536. — — Umkehrbare Isomerisierung von Carotinoiden durch Jod-Katalyse. Ber. dtsch. chem. Ges. **72**, 1340 (1939).

537. ZECHMEISTER, L. and L. WALLCAVE: A Study of Some *cis-trans* Isomeric Dehydro-β-carotenes. J. Amer. Chem. Soc. **75**, 5341 (1953).

538. ZECHMEISTER, L. and F. W. WENT: Some Stereochemical Aspects in Genetics. Nature (London) **162**, 847 (1948).

539. ZELLER, P., F. BADER, H. LINDLAR, M. MONTAVON, P. MÜLLER, R. RÜEGG, G. RYSER, G. SAUCY, S. F. SCHAEREN, U. SCHWIETER, K. STRICKER, R. TAMM, P. ZÜRCHER und O. ISLER: Synthesen in der Carotinoid-Reihe. 13. Mitt. Synthese von Canthaxanthin. Helv. Chim. Acta **42**, 841 (1959).

540. ZIEGLER, H. H. v., C. H. EUGSTER und P. KARRER: Carotinoidsynthesen. XVI. Stereoisomere 1,3,7,12,16,18-Hexaphenyl-ōctadeca-nonaene. Helv. Chim. Acta **38**, 613 (1955).

541. ZSCHEILE, F. P. and B. W. BEADLE: Determination of β-Carotene and Neo-β-carotene with the Visual Spectrophotometer. Ind. Eng. Chem., Analyt. Ed. **14**, 633 (1942).

542. ZSCHEILE, F. P., R. H. HARPER and H. A. NASH: Photochemical Reaction of Iodine with Carotenoids. Arch. Biochemistry **5**, 211 (1944).

543. ZSCHEILE, F. P. and J. W. PORTER: Analytical Methods for Carotenes of *Lycopersicon* Species and Strains. Ind. Eng. Chem., Analyt. Ed. **19**, 47 (1947).

Index of Authors.

(Italicized page numbers refer to the Bibliography.)

Index of Subjects.

(An asterisk * indicates a structural formula.)

Manzsche Buchdruckerei, Wien IX.